If there must be trouble, let it be in my day, that my child may have peace; and this single reflection, well applied, is sufficient to awaken every man to duty…I love the man that can smile in trouble, that can gather strength from distress, and grow brave by reflection. 'Tis the business of little minds to shrink; but he whose heart is firm, and whose conscience approves his conduct, will pursue his principles unto death. - Thomas Paine – *The Crisis, December 23, 1776.*

Contributing Authors

Jerry Adler

Scott C. Barbu

Ben Barnard

Billy J. Bouquet

Doug Cooper

Jack Cotrel

Derek H. Detjen

Toki Endo

Geoff Engels

Jim Farmer

Russell Greer

Jes Hales

Robert O. Harder

Jack Hawley

Gary Henley

Stephen Henley

Dave Hofstadter

Marv Howell

Tom Jones

Rusty Keller

Nick Maier

Ron Poland

Bill Reynolds

Sam Roberts

Ken Schmidt

George Schryer

Pete Seberger

Dave Thomson

Tommy Towery

David R. Volker

Cover Art by Greg Johnson

We Were Crewdogs
V

We Flew
The Heavies

**Edited by
Tommy Towery**

Should the decision be made to publish a future volume of stories such as this and you want to participate, please contact the editor to let your desire be known or visit our web site at:

www.wewerecrewdogs.com

Contact Info:
Tommy Towery
5709 Pecan Trace
Memphis, TN 38135
ttowery@memphis.edu

The cover art was done by Greg Johnson. Greg was born in San Antonio, Texas. He grew up in the shadow of Barksdale AFB, Louisiana, where he developed a great affection for the B-52 and the sound of jet engines. He graduated from Louisiana State University in Shreveport with his BA and after a brief career as a graphic designer he graduated from the Savannah College of Art and Design with his MFA in computer art. After a brief stint working in the television industry he was hired as a professor of computer art and game development at the Savannah College of Art and Design. After thirteen years there he currently teaches courses in a number of 3-D modeling programs and computer languages. His life long passion for aviation is expressed in his paintings. Prints of his artwork are available online at www.GregTheArtist.com.

Table of Contents

Foreword

Chapter 1 -Military Career

Chapter 2 - Cold War

Chapter 3 - Southeast Asia

Chapter 4 - Bar Stories

Chapter 5 - Lest We Forget

Résumé

Heavy Bomber [hev-ee] [bom-er] - *noun* – Any large bomber considered to be relatively heavy, such as a bomber having a gross weight, including bomb load, of 250,000 pounds or more like the B-52.

Foreword
Stephen Henley

When the editor, Tommy Towery, asked me to write this foreword, I was quite honored, and a little apprehensive. I had actually helped my dad as an unattributed editor on some of the stories he wrote for the previous editions of this series. I began thinking about what I should write, and a lot of things came to my mind, but they were all about the stories I remembered from those previous volumes. My thoughts all revolved around the historical and day-to-day roles of the B-52: nuclear deterrence, SAC training and exercise days, sitting alert tours, unearthly tedious inspections, and the pranks of bored crews waiting for the klaxon to sound.

After thinking about it during the final week of my combat readiness training in the B-52H, I realized that this book isn't about the past...well, not specifically. It's about the Crewdogs who crew the mighty BUFF. It's about those ground maintenance crews who keep this warhorse in the air. It's about the sacrifices the families have made through the years: missed birthdays, missed holidays, missed anniversaries, and postponement after postponement of personal plans and goals. It's about those who have made the ultimate sacrifice and paid with their very lives for the freedom of Americans and those whom we desire to free from oppression. It's about hours upon hours of boredom followed by a few seconds of stark terror.

In short, it isn't exclusively about our history; it's about our legacy and our shared experiences, which continue to be faced today. It's about our comrades-in-arms that continue the deterrence of war. To paraphrase from the movie, *"A Few Good Men,"* it's about those who stand up for America and say, "No one is going hurt you—not on my watch," then draw a line in the sand, and look our enemies straight in the eye.

This volume of *"We Were Crewdogs"* includes stories that are serious, funny, sad, joyful, and many varying shades in between. Since the first Gulf War, we have lost three B-52s. Unfortunately, I am writing this on the anniversary of the loss of our latest, Raider - 21. Please keep those families and all the others who have lost loved ones in defense of our country in your thoughts and prayers. Know that their sacrifices are not in vain—we will continue to protect our country.

The Crewdog legacy continues....

i

I was there for Linebacker II (the Eleven Day War). The first night the crews came out to their airplanes and were joking around as usual, and seemed to be excited about their mission. We maintenance folks didn't know yet what was going on. They loaded up, started engines, and taxied out. There were a lot of airplanes taxing out. My ground crew and I took our bread-truck-type van, and drove up next to Charlie Tower. We parked there to watch the show. It was night time, and with all those flashing aircraft beacons, it looked like an interstate highway after a major accident. When the bombers started launching, it was one of the most amazing things I have ever seen. I don't know what take-off timing they were using, but to my perspective, one B-52 would cross the runway hold line, add power, and accelerate down the runway, and before he was half way down, the next B-52 was rolling onto the runway. By the time the first B-52 was lifting off the runway, another one had started down the runway. For almost an hour, there seemed to be three B-52s continuously roaring down the same runway at the same time. Way off in the distance, there was a long line of red rotating beacons stretched out for miles. It was truly an amazing, historic vision. I don't remember a single B-52 aborting its takeoff. Unfortunately, some of those airplanes, and their crews, didn't come back.

By the next night, stories had spread about what was happening, and the crews who came to fly weren't joking around anymore. This was serious stuff now, and we were all worried about the guys who were flying the airplanes we'd prepared for them. It was at that time that I decided I wanted to return to military flying. Flying combat missions, was where I'd felt I was really doing something important. I'd always liked flying military airplanes as part of the crew, and never was particularly happy just working on them. Anytime I saw an airplane landing, or taking off, I had to pause, and watch. I went to the wing personnel office a couple of days later, and put in an official request to be retrained as a gunner on B-52s. Becoming a KC-135 Boom Operator didn't interest me, because they weren't part of the shooting war.

The loss of our flight crewmembers was a sad thing, but all of our efforts, and their sacrifices resulted in both our getting out of Vietnam, and our POWs being released. Our military actions were vindicated in my opinion.

I was accepted into the Gunner's career field, and in late 1973 I completed land survival school at Fairchild, and water survival school at Homestead. In January 1974 I reported to Castle to attend CCTS and become a B-52 Gunner. For me, that was a tough school. Even though I was a trained and experienced aircraft mechanic, I was somewhat unknowledgeable about electrics, and electronics. Even today, it's still a lot of black magic to me. The classes I had to pass on those subjects were difficult, but I managed to get through them. Things like aircraft systems, flight theory, and crew coordination came easy for me. My flying at Castle was in the tail of the B-52F model airplane. Eventually I graduated and returned to Fairchild. This time I was assigned to the 325th Bomb Squadron rather than OMS.

I began to fly as a student gunner while I transitioned from the B-52 F-model into the B-52 G-model. There was a big difference in the two models. In the G-model I got to sit up front with all the other people. I was no longer stuck back in the back of the bus. The fire control system was almost the same, so that presented no difficulty at all. After a couple of flights with an instructor observing, and advising me, I had a flight with an evaluator, and was cleared to fly by myself as a gunner.

I was soon assigned to a crew. In the B-52, we worked together as an integral crew of six people. The gunner was enlisted, and the other five men were commissioned officers. Being a dedicated crew meant you did everything together, sort of like a little family. You flew, stood alert, occasionally went TDY, had classes, and even went on leave at the same time. You did most everything together. Regardless of rank, the pilot (Aircraft Commander) was in charge.

My very first assigned crew consisted of, Captain Garret Cormandy (AC), Lieutenant Page Wagner (CP), Captain Steve Issac (RN), Lieutenant Rich Nissing (N), Lieutenant Charlie Micheletti (EW), and myself. It was a great crew. We got along well, and worked efficiently, and professionally together. One interesting aspect of this crew was the Navigator, Lieutenant Nissing. He had been a gunner prior to being commissioned, and had flown some combat in the B-52. He was always willing to advise me if I had a problem that needed help from a gunner's perspective. Occasionally, when we had fighter attacks against us, he would come up and be a gunner for a little while. I was with that crew for over a year, so we got to know each other very well. I believe he also had an identical twin brother who was a B-52 pilot at a different base.

A word here on being on a B-52 crew from an enlisted man's perspective. Almost anywhere else in the military, enlisted personal do not mix with officers. There is a social standing that is both written and unwritten that separates them. On a B-52 crew, you were forced by circumstances to exist on a much closer level. The enlisted man flies with them, eats with them, travels with them, occasionally sleeps with them, goes hunting and fishing with them, parties with them, and some even died with them. The officers on his crew become his closest friends. I always gave my officers good, proper military respect by calling them by rank, or rank and last name when in uniform, especially if there were other higher ranking officers around In the airplane, we addressed each other by crew position. Out of uniform, I could loosen up a bit and use first names, but even then, should a colonel perhaps be nearby, I'd use the formal address. I saw no reason to cause one of my officers to get chewed out by a superior because of my actions, and embarrass him as a result. In a normal Air Force squadron, you might find 10 to 20 officers, and perhaps 300 enlisted. In a B-52 squadron, you might have 20 or so enlisted and 100 officers. The dynamics of that situation were certainly flipped over, and could occasionally add to the confusion a bit. The Tanker crews had the same situation, but with fewer officers.

Other than flying, our main responsibility was to pull alert, and be ready to fly an EWO mission if called upon to do so. The American people don't know it, but I would say that those of us who were part of the TRIAD, the missile guys in the silos, the submariners under the sea, and the flyers of the B-52s had the most responsible, and important job there ever was. If we hadn't been there, and ready to do the unthinkable, perhaps our country wouldn't even exist as it does today.

I pulled lots of alert, waiting for the notice that would send me along with everyone else off on a nuclear mission. Alert was a lot like going to jail for a week, every three weeks. Getting qualified to go on alert was probably the tensest and most nerve racking thing I ever did. I'd almost rather fly a combat mission that go through an EWO certification before the Wing Commander and everyone else around with silver leaves on their shoulders. We were like fresh meat to a pack of wolves! Whew! To stand up there before all that brass, and brief them on the entire mission from your own perspective was very stressful. Then being interrogated by each one of them in turn was awful. You just knew that while you were talking, or answering questions, you were either doing very well, or you were hanging yourself. My first certification was scary, and all the rest over the years never got much easier. You could often hear a crewmember's voice rapidly go up and up in pitch while they were briefing their part of the mission.

While I was on my first crew at Fairchild, I stood a lot of alert there. We also stood satellite alert at Glasgow AFB, Montana. If we were on alert at Fairchild, it was from Thursday to Thursday. If at Glasgow, it was Wednesday to Wednesday. To pull alert at Glasgow, an Air Guard C-130 would fly us over and return the off-going crews. The following Wednesday, they'd take us home after we'd been relieved. I went up to the C-130's flight deck one time on a ride back to Fairchild. The Nav was a Major, the Engineer a C/MSgt, the Copilot a full Colonel, and the Pilot was a One Star. I didn't stay around to visit.

Glasgow was interesting because the base itself had been closed, so the alert facility was about all the military that was still there. At Glasgow we didn't have any ground training, no simulators, and no staff to harass us. It was like a vacation from all that other stuff. On average we'd have one simple alert exercise per week I remember us having an elephant walk only once. I may be wrong, but I think to have a moving exercise, either the Wing Commander, or his Deputy, had to be present. The old military buildings and housing had all been turned over to civilian uses. One of those civilian uses was a national school for unwed mothers.

We had a standard underground/above ground alert building, but also had some outside modular quarters for crews near their airplanes. Summers were nice there, but winters were awfully cold. I remember one tour when, for a couple of days, the crews in the mods weren't allowed to leave them - even to go eat chow in the main facility. For those couple of days, the temp/chill factor stayed down around -80 degrees. The guys out there in the mods got a little bit hungry.

One night in the main facility, the crews were playing cards (Bouree probably). A pilot I think was named; "Major Best" was playing. He was of short stature, so he was called, "Shorty Best." He called over to his mod where someone had a popcorn machine. He told

George Hitzler, his gunner, to bring some popcorn over to the game. Over at the mod, someone had found a frozen, very dead, duck lying outside, so they put it in the bottom of a large paper sack, and topped it off with popcorn. The bag was delivered to the game and for the next hour or so, everyone took turns digging a hand down into the popcorn. That kept it up until someone found feathers.

My crew was the last crew to leave Glasgow when they discontinued satellite alert there in 1975. We were off alert, but in the blue alert truck, at the local bowling alley when a bunch of young girls (perhaps some of those unwed mothers in their late teens to early 20s) invited us to go skinny dipping out at someplace call Porcupine Creek. For some reason, we never made it there. Don't know why. Later, we borrowed a car from the mess cook, and went to town and had a few drinks. The next day we flew the last B-52 post-alert sortie out of Glasgow when we headed back to Fairchild.

The same crew took an airplane to Tinker AFB in Oklahoma City once, and after checking into billeting, we all went to the Officer's Club. That place was completely dead, with only a couple of people in it. The guys asked me to take them to the NCO Club, so that's were we went next. When we got there, it was packed with guys and girls. There was a small stage, and there was a lady stripper up there doing her thing. Shortly after that, a fight broke out and the MPs showed up to restore order. We had a ball, and all the guys on my crew kept telling me that NCO clubs were much better that the O'Clubs. We ended up in a bar downtown.

My next crew after that first one had two copilots on it. At that time several crews at Fairchild had two copilots. One of my copilots was a Lieutenant named Jack Evans. That was in 1975, and I ran into him again in 1990 in Guam. Just before they deactivated the B-52G squadron there, and prior to Desert Shield, he took over as the Wing Commander. He was a full Colonel then.

I don't remember the name of one mission I was on out of Fairchild, but it was scheduled to be a long, 17-hour flight, returning to Fairchild. During mission planning I got a menu from the in-flight kitchen, and it showed that they offered frozen meals (sort of like little TV dinners). I called B-52 Job Control, and found out that they had ovens that could be put into the airplane if we desired it. I talked it

9

over with the crew, and we all ordered frozen dinners, so I ordered an oven. An hour or so after takeoff, someone wanted to try a frozen dinner, so I put it into the oven, and turned the rheostat to turn it on. There was a loud pop, and several circuit breakers on various C/B panels popped out. We reset the breakers and tried it again, but the same thing happened, so we gave up on the idea of having a nice hot meal. I think I ate mine, but it was still frozen when I did. That was my longest B-52 flight, and I don't envy the guys who are flying even longer ones today.

We were taking off on a scheduled, 12-hour Snow Time mission one morning out of Fairchild. We were number three in a five ship MITO takeoff with, "like aircraft, 15 second takeoff intervals." On the interphone, I heard, "coming up on un-stick. Ready, ready, now. Kaboom! Our number five engine exploded. The airplane was really rocking and rolling as the pilot fought to control it with the loss of thrust on the right side, plus the wake turbulence from the two airplanes in front of us hampering his efforts. He saved the airplane; we stabilized in a climb, and tried to evaluate the problem. The EW put the sextant up to take a look at the wing, but it was clear. We didn't have a fire, and no apparent damage. We aborted the mission, but due to our heavy fuel load, we ended up flying around for almost five and a half hours until we were light enough to land. I never understood why Boeing hadn't incorporated a fuel dump system on the B-52. It's such a simple system, with fuel lines running to a simple dump manifold where fuel is allowed to flow out of the airplane, and quickly get the weight down to a manageable level. Passenger airplanes are all required to have them, and so does the KC-135. Flying around for hours in a sick, dangerous, damaged airplane to burn-off enough fuel so you can land always seemed to me to be the result of some lousy engineering.

I was the Gunner, and Captain Gordon Hallgren was the EW on my crew when we had the UFO incident as he described in; "*We Were Crewdogs I*." My version of the incident differs a little bit from his, but I was there, and we definitely had something strange and unusual happen. I rode along on two hard, scary landings at Fairchild in the fog. On one of them, after landing, we found a taxiway, pulled onto it, and stopped, then requested a follow-me truck. In the dense fog, it took him almost 45 minutes to find us, and lead us to parking. I saw Major Hallgren again several years later, when I was at Andersen, and he came in with the IG. On alert one time at Fairchild, Captain Hallgren had the alert truck, and was gone to school working on his Masters, I believe. We had an alert, and when we got to the airplane, and started engines, we had a wet start on one of them, that caught fire. The pilots were maintaining the start, and trying to blow the fire out, which I guess was spectacular. Unknown to us, during the fire, he arrived but waited outside in the truck until the fire was finally out. He then joined the rest of us inside. As he came up the ladder next to me, he leaned over and said, "I don't know Rusty, I must have been trained wrong, but I just can't see being in a big hurry to get into something that's on fire!"

Another of my favorite EW's was Captain Rich Colarco (sp?). You old EW's might relate to this, but on a flight one day, he'd been doing all his radio calls. He then cleared off interphone to heat some coffee in the hot cup. When it was hot, he plugged into the cord from the DI's interphone panel. I heard him in my headphones say, "I've got some hot coffee, anybody want some?" Silence. "Hey, I've got some hot coffee, anybody want some?" Then we all heard, "Fort Worth Center will take a cup with cream and sugar." Rich was a blur of motion getting back to the DI's seat, and resetting the radio selector. Years later when I was a C-130 Engineer, I ran into him at RAF Mildenhall in England. He was a Major, and a Compartment Commander on AWACS Aircraft.

Going on alert one day at Grand Forks, my Radar Navigator, Captain Tommy Taylor showed up with six bright yellow toy fireman type helmets, complete with red rotating beacons on top, and sirens. He'd bought them somewhere, and before we went into the assumption of alert briefing, he gave us each one of them. We waited until everyone was in their seats, then in column; we all walked in with lights flashing, and sirens wailing. It was so funny. Everyone in there cracked up. The really funny thing was later, when a couple of newbies were convinced by their seniors that those helmets were issued for alert ground running responses to the klaxon, and they'd better not get caught without theirs. They spent quite a while trying to get them issued; until they found out they'd been had.

I remember flying during the Super Bowl one time, and it was funny because every time an airliner would check in with Center, they'd ask about the score. The controllers must have had a TV in there that they were watching while they directed traffic. We'd hear something like; "Denver Center, United Flight 406 with you, flight level 350, what's the score." "United 406, radar contact, it's 7 to 10, Cowboys, first quarter."

A copilot on another crew in my flight had a wife who baked him a big tin of chocolate chip cookies to take with him on alert. He soon noticed that his cookies were disappearing faster that he was eating them. That happened every time he went on alert. Finally, he had his wife make them with a chocolate laxative added. Surprisingly, on his very next alert tour, his navigator came down with some sort of intestinal problems resulting in serious diarrhea. It drove the Flight Surgeon crazy trying different medications, attempting to get that Nav's stomach problems cured. For some reason he was afflicted with those problems only when he was on alert. For several months that continued, and I don't think that Nav ever figured it out. Stolen cookies, you know, might be bad for you?

I flew a Red Flag Mission once out of Grand Forks. My pilot was Captain Bob Johnson. We were low level, and had an F-4 on our tail, and were right down in the dirt, when we flew over a ridgeline with 75ft showing on the radar altimeter (the copilot told me later). We landed afterwards at Castle and hopped on a KC-135A and flew to Nellis AFB for a debriefing. There, I talked to a fighter pilot Major who described the intercept to me. He said, "You guys were blowing dirt like a super grass-blower, and I was closing on you, but then I climbed up a couple hundred feet, and watched you guys taxi up one side, and down the other side of a mountain, then I closed, and shot you down." Captain Johnson," had been a T-38 IP in UPT at Williams AFB. He eventually got an assignment to fly the U-2 (TR-1). The last I heard he was the Ops Officer in the U-2 squadron at Beale AFB, California.

While pulling alert at Grand Forks, there was a tanker Navigator that was considered to be sort of a screw-up. His reputation was that he would often do strange, stupid things in his normal life, not as a joke. One day, I was in my crew's modular quarters which were right behind our airplane. We were sortie number one, which was the furthest parking spot from the main facility. The wind was blowing really hard, directly from my B-52H to the facility. The rest of my crew was at the facility, and they had the truck which they'd parked outside of it. The before mentioned tanker Nav, had sat down in our truck because his crew was gone with his truck. He was sitting there behind the steering wheel reading a book, when the Klaxon went off. He started our truck, and took off for his tanker at the other end of the line. My crew came running out, and there was no truck there for them so they had to run all the way to the airplane. I knew where they were, so when the Klaxon went off, I strolled out to the airplane, and met the Crew Chief at the entry control point. We went in together (no-lone Zone area), and started removing engine covers, pitot tube covers, ground wires, etc. While doing that, I could see my crew running towards me. There wasn't any snow on the ground, but it was cold, so they were all wearing parkas, and running directly into the face of a strong headwind. They seemed to be running flat out, but not making much headway. Poor guys, they were exhausted by the time they got to the airplane. A certain tanker Nav, and his pilot got to meet a very angry B-52 pilot (mine), when it was over.

I was on alert at Andersen on D-models when we had a Klaxon just after 4 AM. We never had horns in the middle of the night, so we were all somewhat sleepily pissed off. We raced our trucks to the alert line which in Guam was a long ways away. I parked the truck, and I got myself up into the tail, into my seat, and pulled the seatback up against me. Engine-start cartridges were blowing off, and gray smoke from them was turned yellow under the yellow lights of the alert area. Everything was getting very noisy with the continuous swoosh of the cartridges firing. I put on my headset and heard all the various sorties asking Command Post to; "say message." Finally I heard "Message follows," then all the letters and numbers for the message. Shortly after that, I heard my Radar in a very high squeaky voice say, "Pilot, I have to come up and open the box." "What," said the pilot! The Radar repeated it, and was told to go ahead and do it. You see, we never opened the box. To open the box which held the documents and authenticator tickets was a real nuclear war thing to do. We just never did that. I cussed out loud, yanked off my headset, and put on my helmet. Then I wiggled my way into my parachute harness because I, along with everyone else was sure we were going flying, and there was only one place we'd be going. Somehow, back in Cheyenne Mountain the computers had decided that the U.S. was under nuclear attack and blew the whistle to launch a retaliatory strike. We got awakened at 4 AM in Guam. The big boys at SAC finally figured out what had happened and advised our Command Post as well as all the rest of them I suppose. Command Post called everyone and said, "Assume normal alert." Once again, I heard lots of, "say message." Turned out, there was no message to get to normal alert from where we were. That had never happened before. It took awhile for Command Post to finally convince everyone to assume normal alert even though there wasn't a message to do so. It certainly scared the Hell out of everybody that morning.

Everyone who ever flew an ORI knows that the airplanes were often pretty sick, and really shouldn't have been in the air, much less pushed all the way through a mission. One night out of Guam we took off in our B-52D on an ORI mission. The airplane wasn't feeling very good, and lots of things started going wrong with it, mostly electrics and hydraulics. We flew though. We'd accomplished everything but the bomb run, and while doing that, we were out over the dark Pacific Ocean. It was very dark, with no apparent horizon other than the lights of Saipan a few miles away on one side. Suddenly the Nav yelled, "Climb, climb, climb." I was told later that he'd caught us in a descent only a hundred feet from the water. Perhaps five seconds from eternity. As soon as our last bomb release, we declared an emergency and climbed out towards Guam. During our pre-landing checklist, the pilot asked for Gear Down. I felt the gear start to cycle, then heard the copilot say, "That's not funny!" It seems they had an unsafe gear indication on the forward trucks. We tried everything we, and the command post, could think of then did a low fly over. The SOF ran his truck at max speed down the runway with a bright spot light on us trying to see if the gear was down or not. He said they all looked down and locked, so down we came. When we touched down, it was a very unusual landing. I never felt one quite like it, but wouldn't be able to describe it adequately. It was sort of a jerking, rolling, wiggling landing. We stopped on the runway because the pilots didn't have any steering. I got out through the aft wheel well and walked forward. Everyone was standing there looking at our two front landing gear trucks which were pointing directly away from each other. The left one was pointing 90 degrees to the left, and the right one was pointing 90 degrees to the right. No wonder we had no steering. You might even say we had our auto-braking system on, as the tires, 90 degrees to the direction of travel, didn't roll very well.

Well anyway, I was a gunner at Fairchild on the B-52G starting in 1974, then transferred to the B-52H at Grand Forks, AFB, North Dakota, in 1976. I became an Instructor Gunner on the H-model, and was there until 1979 when I transferred to Andersen AFB, Guam. I flew B-52D's there until mid-1982 when I was released from the B-52 world, and given orders to become a Flight Engineer on the C-130H at Dyess AFB, Texas. I finished up my B-52 career with just over 2,100 total flight hours in various models of the B-52. I left SAC for MAC, and went to Flight Engineer Performance School at Altus AFB, Oklahoma. I trained on the C-130E at Little Rock AFB, Arkansas. From my home base of Dyess AFB, I flew the C-130H all over the world for four years, and logged about 1,400 hours. I retired from active duty USAF in 1986 as a Master Sergeant (E-7).

After my military service, I flew little airplanes around the country for awhile as a Commercial Pilot, and then in 1989, I took a civilian job as a console operator on the new B-52G WST simulator in Guam. I trained for that job at Castle AFB, and eventually qualified to run both the console on the pilot's, and the EW/FCO's defensive station consoles. In 1990 we operated during Desert Shield, and trained the B-52G crews that eventually flew Desert Storm missions. After that, I worked various jobs, among them: being a professional Scuba diver (Dive Master), commercial pilot, horseshoer (farrier), and driving a stagecoach pulled by mules and horses around Tombstone, Arizona. For the last few years I worked on heavy jet airliners as an aircraft mechanic. I'm retired now and do a little bit of writing (more as a hobby than a livelihood). I've had several magazine articles published, and have written a couple of self-published books. I'd say the main highlights of my life, are my wife Sharon, my three daughters, and my five granddaughters. Imagine that: Me a grandfather!

To all the old B-52 gang - Hey!

To all the old Gunner's - C'est La Vie!

A Crewdog's Tale
Dave Thomson

I recently had a reunion with some of the crewmembers I flew with in SAC about 35 years ago. They had been contributors to *"We Were Crewdogs II and III."* It brought about a flood of memories that I hadn't thought about for many years. I don't know if time has been kind to my mind in the accurate recall of details from that far back but I'll try, and beg forgiveness for errors.

I graduated from college in January of 1969, knew my student deferment was gone and the draft would get me sooner or later. I figured it would be better to get my obligation done and then go out and make my fortune in business. The Air Force seemed more appealing than the Army or Marines, slogging through the jungles of Vietnam. I took all the tests and found I was well qualified but had one problem. I was born in Canada of US parents and had dual citizenship. To be an officer, I had to be a US citizen and dual doesn't count. While I was going through a lengthy court process with Immigration and Naturalization Service (INS), my greetings came from the President. I rushed to my recruiter and he advised me to sign up for the delayed enlistment program that would ensure my entry to the Air Force. It would buy me 120 days and I should have my Officer Training School Class assignment by then.

I got my INS papers completed, submitted all the applications, and waited...and waited...and waited. My deferment expired and I had to ship out for basic training without hearing from the selection board. I was an Airman Basic going through training at Lackland AFB. When graduation came six weeks later, I got orders to go to Chanute AFB to become a personnel specialist. I was standing in line (a common occurrence in Basic) to get on the bus to Chanute when the Drill Instructor hollered to me that I was in the wrong line. I'm sure he enjoyed my bewilderment until he revealed that my orders had been changed; I was accepted to OTS with follow on training as a Navigator.

It was another six weeks until my class date so I was put in casual status at Lackland which meant some sort of menial duty to keep me out of trouble. My task was to maintain the training records of the officer candidates in the OTS class ahead of me, my soon-to-be upper classmen. I found it very interesting to look at the Air Force Officer Qualification Test (AFOQT) scores of the pilot selectees. There are three aptitude scores: Pilot, Navigator, and Officer. Of a possible 95 points, mine were 55, 95, and 90. I had passed my flight physical with flying colors but my recruiter suggested that I apply for OTS with nav first and pilot second to increase my chances of being selected. I did and was selected as a nav. Thinking later, I believe the recruiter had a navigator goal to fill and I was it. Many of the students who were selected as pilot had scores lower than mine and some had 35 as their highest score in any category. Later in my career, I found that it is very much a pilot's Air Force. I served with a lot of outstanding pilots but kind of wonder about the "35s" out there.

I got my butter bar from OTS November 13th and had a class start date three days later for Undergraduate Navigator Training (UNT) at Mather AFB. Training went well and I had a good variety of assignment choices and selected Nav Bomb Training (NBT) because it would put a navigator in a central mission role. I remember being surprised at UNT graduation that my parents flew out from Wisconsin. My mother pinned my navigator wings on and my dad gave me a new watch. It was one of those newfangled self winding watches. You didn't have to pull out the winding stem and wind it every day, it did it for you. Not batteries and digital like today but a system of counterbalanced weights that wound it with the motion of your arm.

I didn't have much time between UNT and NBT schools, just three weeks. Personnel notified me that there was a survival school slot at Fairchild AFB open for that same three weeks if I wanted it. I agreed. I remember coming out of the grueling field training one day, getting my training certificate the next, catching a plane and starting NBT the next day. I did well in NBT, third in my class, which meant I got third pick of all the assignments available. During the later stages of training, I had long discussions with a buddy about the pros and cons of various aircraft and was hoping for an FB-111. Well, two FB-111s came down to the class and Number One took the first one and my buddy took the second. I had my choice of all the remaining B-52 assignments. I knew the D-models bore the brunt of the Arc Light assignments so I picked Carswell to put my training to use in probable combat in Southeast Asia (SEA). This meant more schooling, this time down the road at Castle AFB for B-52D crew training. Upon successful completion of that school, I found myself driving a Volkswagen Beetle with all my worldly goods in it across the Southwest desert in late summer enroute to Fort Worth. Still not done with training, I started local area training and passed my check ride in December as a combat ready navigator. I had been in training for 2½ years and was finally qualified to fly without an instructor over my shoulder.

The next task was to put me on a crew. I was in the 20th Bomb Squadron that had an upcoming tasking to fill an Arc Light deployment. A Stan/Eval crew was tasked for the deployment but the navigator refused, saying he was a conscientious objector. He did not object to standing alert with a tasking to drop nuclear weapons on the Soviet Union because that was going to be the end of the world anyway, but he objected to dropping iron bombs in a conventional war we didn't belong in. He was jerked out of flight duty and became commissary officer until they could figure out what to do with him. What nav was a SEA volunteer and available? That new guy, 1Lt Thomson, but the Stan/Eval job didn't go with it - you're just a body for the deployment.

It was still several months before the deployment so I certified the war plan and started pulling back-to-back alerts and flying training sorties with the crew: Maj Fred Green (P), Capt Joe Rea (pronounced Ray) Fontes (CP), Lt Col Dick Kunz (RN), Maj Matt Smith (EWO), and SMSgt Joe Adams (G). Carswell may have been a little different than other bases because we were expanding into two bomb squadrons and had a lot of people on alert. SAC was big on accountability for many things including management. One of the management measurements of quality of life was minimizing back-to-back alerts, so Carswell had two changeover days - Thursday and Friday. You may get off alert on Friday and go back on the next Thursday or Friday and that was a back-to-back tour and reportable to HQ. But, if you got off on Thursday and went back on Friday the next week, then that was not back-to-back and not reportable.

I was on alert with the crew early February of 1972 and preparing to fly a First Sortie After Ground Alert (FSAGA). A FSAGA mission was very high on the accountability list to HQ SAC. The plane sits alert for a month then the nuclear weapons are downloaded, the airplane is uncocked and taken off the alert line but nothing else is done to the plane. The crew coming off alert then flies a simulated combat profile and the results are sent to HQ SAC, measuring whether that plane and crew would have been successful on an actual wartime mission. It was to be a night mission and everything was normal through taxi to the pre-takeoff position. An on-time takeoff is one of the critical events. We announced to the tower that we were ready for takeoff but were told to hold our position. We were sure the tower personnel were aware we were a FSAGA mission but as a discrete reminder asked anyway if command post was aware we were being held. We got a roger. We sat for a while, aware that up to 15 minutes after scheduled takeoff still met on-time takeoff criteria. With about a minute left in that window, we asked again for takeoff clearance but were told to hold. We sat for about 20 minutes more then tower said to contact the command post for words. Command Post said the mission was cancelled, return to parking and come into the command post. When we got there, the wing DO said go home, go into crew rest, report back at 1500 hours, we are being deployed, he couldn't say where or how long.

I didn't know what to make of that. I knew there were a wide variety of global missions SAC could task us for including SEA. At

1500 I showed up with all my alert/deployment gear including parka and mukluks and was handed a flight plan to Guam. We were to be in the first wave of something called "Bullet Shot." The orders were for 179 days and didn't give a destination, just something about needs of the Air Force.

As a new Nav, your first flight out of CONUS adds a new level of trepidation to the flight. I still remember that we were out of sight of any land for many hours before our route took us over Wake Island. I looked at the radar where I hoped the island would be but there was only a bunch of weather out there. We flew on, the feeling growing that I was lost. Finally, we got close enough to tell that there was an island under one of the thunderstorms and we were within two miles of where we were supposed to be. The arrival at Guam was uneventful other than the shock of stepping off the aircraft into a tropical climate in February. We were quickly oriented and worked into the schedule. Within one week of sitting alert at Carswell AFB and preflighting nuclear weapons, I was dropping iron bombs in SEA.

Because we were a Stan/Eval crew with numerous Arc Light tours (except for me and the CP), we were lead on my first combat mission. I was a greenhorn but I couldn't have been with a better crew to cover my butt. I don't remember making any critical mistakes but that may have been selective memory. Being lead on every mission was definitely challenging for a navigator and probably helped me mature in the job faster than my contemporaries.

We had at least a dozen missions together by Easter of 1972 and had rotated to U-Tapao. Intel had been telling us that there were no SAMs south of the DMZ. However, the day before, a cell reported a possible SAM launch at a target just south of the DMZ but Intel discounted it as a probable rocket. Another cell an hour later in the same area reported two possible SAMs but those were again discounted as rockets. The third cell in the area reported three launches and that got Intel's attention. They diverted all cells for the next 12 hours while they evaluated the data. We were lead of the next cell that was targeted near the DMZ. Intel briefed us that there was a major NVA incursion across the DMZ and it was possible there may be SAMs in the vicinity. Being forewarned, the crew was extra vigilant.

The pilot told us later that it was a relatively clear night with low clouds at the end of the winter monsoon season. The bomb run was going smoothly through bomb release and post target turn. The CP was watching something going on below. There was some kind of strobe light under the clouds. It pulsed half a dozen times, paused and pulsed a half dozen times more. When Rea realized that it was moving, reality struck. SAMs! All he could do was point with his right hand and reach his left hand across the center console, hit the pilot's shoulder repeatedly and say "Ah…Ah…Ah." Of course this maneuver doesn't leave any hands free to hit a microphone switch but Fred recognized the situation and announced to the cell "Orange Cell, SAMs two o'clock low, continue break left and maneuver." The countermeasures were successful and the closest SAM was probably a half mile away. What was pieced together later was that Rea had watched the launch of six SA-2s, their first stage separation and the ignition of the second stage. The ECM prevented guidance and they just went ballistic. When we got back to debriefing, Intel finally acknowledged that there were definitely SAMs south of the DMZ.

Another mission started out relatively uneventfully. We did our usual pre-mission preparations followed by the crew brief. All crews in the three-ship cell formation sat through a formal briefing that started with a time "check," not a time "hack." The only time the word "hack" was used was to release bombs. There was a story (probably untrue) that a hapless aviator once asked for a time hack over a tactical frequency and bombs fell all over Vietnam. As usual, we were lead so all the mission timing fell on me. We had a time over target (TOT) that had to be hit plus or minus three minutes. That target time was de-conflicted with ground forces in the area and other air operations so there would be no mishaps. If you could not make your TOT, you withheld and went to a secondary target. Takeoff time was scheduled with some slack to accommodate launch problems so there was a timing box in safe territory where cells could maneuver freely to adjust timing. If you were late, you could scream straight to the exit point or if you were early, you could make big S-turns or even orbit. Once you left the timing box, you were committed to the preplanned mission corridor and little flexibility to adjust timing.

On this mission, I had calculated the exact time I needed to exit the timing box and had updated it with actual winds at altitude. We hit the exit point of the timing box about 30 seconds early with numbers two and three in trail maintaining their prescribed positions with station keeping radar. We were in non-hostile airspace about 40 minutes to the

TOT and hadn't got to the pre-IP so I had the pilot reduce airspeed 20 knots. A little while later I recalculated my estimated TOT and we were then a whole minute early. Had we picked up an unexpected tailwind? There was a left turn coming up so I maneuvered the cell to the right side of the corridor and took the wide side of the turn, guaranteed to lose a minute. Except, my next update said we were two minutes early and running out of time and space to adjust. I was able to do a couple s-turns in the corridor up to the IP and the crews behind me were probably going crazy trying to maintain position. We got to the IP and had to run the prescribed airspeed and headings for a ground directed MSQ ground radar release. Even though I calculated we would be 2-1/2 minutes early, we were still in the bomb release criteria. At the release "hack", I marked the actual release and ran the appropriate checklists. In the post-target procedures, lead always announces the actual release time to two and three so all the paperwork agrees. When I announced the TOT, I got an immediate response from the cell mates that we were actually almost six minutes later. I started to argue but then asked for a time check from my crew. My watch and theirs were almost six minutes different. Then I was really confused. I knew we all set our watches correctly. I deferred to the RN's accounting of the actual release time. When everything was recalculated, our actual TOT was three minutes and five seconds late. It was outside the bomb release criteria and I knew we were going to be severely reprimanded for it. Although it was my fault, I was dismayed that neither of the other Navs in the cell realized I was running late or questioned my odd maneuvers. The answer finally came to me about 20 minutes later when my watch stopped. We weren't gaining time, the watch my dad gave me at Nav school graduation was slowing down and finally stopped. The tropical conditions and all the sweat had rusted the neat self winding mechanism and it ground to a halt. Fortunately, the wing staff never noticed the TOT discrepancy or chose to overlook it. Needless to say, I bought all the beer at post flight.

As a post script to this adventure, the pilot set up a challenge and I accepted. We had a bet that if I could hit the TOT plus or minus five seconds, the crew would all buy me beer at the end of the mission. If I missed the TOT by more than five seconds, I would buy the beer for the whole crew. I drank more free beer than I had to pay for.

As that tour wore on, it became obvious to SAC planners that the massive commitment of aircraft and crews to the SEA conflict was not going to end soon and all the crews were deployed with 179 day orders.

If anyone went over 180 days, that would be considered a SEA PCS tour and all the rules changed. To prevent that, they had to set up a rotation schedule. Because we were one of the early crews in Bullet Shot, we got curtailed at four months and sent back to the states for a 28-day break. The next four tours were all rotated at about five months.

There was an airplane that needed to be ferried back to Barksdale AFB for heavy maintenance and we were selected. Knowing that we had our own ride, we stocked up on all the heavy treasures that aircrews accumulated from the Pacific exchange systems. We had multiple Hibachi pots, papasan chairs, quadraphonic sound systems with reel-to-reel tape decks and gigantic speakers, carved mahogany room dividers, etc. A rack was slung in the forward bomb bay with plywood flooring and all the loot secured on it. We then headed back to the states with one refueling about 300 miles east of Guam and a second about 300 miles off the California coast. The first was a tanker that launched with us and the second was a standard point-parallel rendezvous with a tanker out of the states. Other than all the combat missions from Guam, this was my second long trip navigating the Pacific. I had more confidence in my navigating abilities than the first trip but started questioning things when we approached the second refueling point and couldn't find the tanker. No refueling meant diverting into March AFB. We were out of range of any land radar returns but established radio contact and confirmed his beacon was on. The only beacon I found was 150 miles north of the rendezvous point. He had the gas and we needed it so he turned southwest and we turned northeast to intercept. I don't think that rendezvous was in any procedure manual but we got rolled out under the tanker and got the gas. While refueling, the tanker has the navigation responsibility which included penetrating the Air Defense Identification Zone at 200 miles. The tanker made a large turn to the southeast while we were on the hose and successfully got us through the ADIZ without getting fighters scrambled.

When we got into Barksdale, we found that Carswell had sent a T-29 to retrieve us and a Dyess crew. The T-29 was a twin engine propeller plane used for UNT and NBT and in this case a logistic support plane for the wing. It was not designed to carry 12 B-52 crewmen and tons of swag from the Pacific. We added up all the weight and the poor pilot determined that the required takeoff speed was above the flap limitation and it would require all 12,000' of the Barksdale runway in the summer heat. The pilot had never made a no-flap takeoff but would give it a try. We crammed everybody and everything into the plane and took off. I believe there was discussion later about how little we cleared the barrier by at the end of the runway.

I don't know what the circumstances were but on the next tour, Maj Matt Smith was replaced on the crew with 1Lt Dave Hofstadter (EW). I also found that I was in the 9th Bomb Squadron then and we were crew S-51.

We got back to Guam in time for a typhoon evacuation. It turned out to be a false alarm but some crews had been diverted into Okinawa. The Japanese were not happy to have bombers on their island. We were sent on a KC-97 to Okinawa to round the crews up out of the bars and bathhouses and return them to Guam. Most of the herding had been done by the time we got there so it was a quick turnaround. We got back into the schedule and resumed combat missions. Our pilot, Fred Green had worked a deal that got us moved to U-Tapao and involved with the Pave BUFF testing program. Dave Hofstadter wrote an excellent account of our participation in the Pave BUFF program in *"We Were Crewdogs II"* but that's not the whole story.

All of the Pave BUFF missions were flown out of U-Tapao and the test was to have been concluded about Thanksgiving - the same time as our scheduled crew rotation. There had been some maintenance or weather cancels that pushed the completion back to early December. The staff left the decision up to us whether we would depart on our regular rotation or volunteer to extend a couple weeks to complete the test. Being the astute Crewdogs that we were, we could add 28 days to late November and see we returned before Christmas, or by "volunteering" to extend and complete the test, we would return after Christmas. We violated the adage that you don't volunteer for anything.

As I remember, the head of the Litton contractor team decided to throw a party to celebrate the successful completion of the test program. He was staying at a nice house at Pattaya Beach, the tourist haven on the Gulf of Siam north of U-Tapao. All of the crews and contractor personnel involved in the program were invited. A group of Thai ladies were cooking and serving an excellent variety of Thai dishes. Everyone was having a great time there and the drinks were flowing freely. However, our copilot Rea normally had about a two beer limit. He had unwisely pushed through the limit and was working on his fourth beer when he noticed one of the Thai women. She was beautiful and extraordinarily well developed in the chest area. Rea ogled her for a while then slid over to our host and asked him how to say "Nice tits" in Thai. Without batting an eye, the host said you can say "mak mak nom" or "suay mak nom" but my wife is used to either one.

Rea was almost invisible the rest of the night. We rode a Bhat bus back to the base, realizing that it was way too short a time to squeeze in 12 hours between bottle and throttle for our combat mission in the morning. But, like the professional Crewdogs that we were, we made the mission without incident.

On December 5th, 1972, we departed U-Tapao for our 28 days stateside. Little did we know what would happen in the next few weeks. I had accumulated over 100 missions by that time and the pilot and gunner were over 300. In all that time, I had felt frustrated that we were bombing seemingly insignificant targets along the Ho Chi Minh Trail and South Vietnam. I remember the RN looking at target areas through the optical bomb site and saying it looked like a moonscape, barren and cratered from prior strikes. Where we needed to go to put an end to the war was Hanoi and Haiphong. I realized that Hanoi was the most heavily defended city outside of Moscow at the time but I was ready to accept the risk. Cut off the snake's head and the rest will die. We were being politically constrained; it wasn't the military's decision how to prosecute the war. I had mixed emotions when I visited my parents in Milwaukee at Christmas and heard the news of B-52s being shot down in raids over the North. I was frustrated that I missed the action but thankful that I wasn't shot down.

Our return to theater was a typical Young Tiger logistics trip. We caught one KC-135 logistic flight to March AFB then stayed overnight waiting for our next flight to Guam. We loaded all out gear in the back of a truck that took it out to the plane while we waited in base ops. The plane was carrying max capacity and all the gear was floor-loaded in the middle of the plane between plywood walls that left 18 inches between the edge of the folding canvas troop seats and the wall of baggage. All the passengers were then squeezed into the remaining space. The flight to Hawaii was uneventful, one person terminated there and the rest continued on to Guam, after the plane was refueled. After landing, all the baggage was brought into base ops and retrieved by the owners. My nav bag with all my navigating equipment was missing. It was the standard issue black briefcase with my dash-1, checklist, celestial manuals, protractors, dividers, etc. After confirming that nobody else had accidentally picked up my bag, I reported it missing and had to get a reissue of everything before flying. Nobody could explain the disappearing act.

We got back in theater before January but the Linebacker II raids over the north were over. I think the third tour included the cease fire and Paris Peace Talks. There were still combat missions being flown in Cambodia for a while but I remember when the combat missions ceased, there were more aircraft than parking spaces. One of the parallel runways at Anderson AFB was shut down and planes were parked nose to tail on the runway.

One commendation I have to add here is that the EW, Dave Hofstadter, and I served four tours together. I was never fired on while Dave was on the flight. We were shot at when we had a substitute EW on board and I was shot at when I flew with another crew as DNIF cover but never with Dave in the ECM suite. Some may attribute it to good luck but I think it was superior skill. I think Dave ended up with 118 combat missions and I had 175.

That tour was five months and ended up at U-Tapao. We had the usual Young Tiger redeployment for our 28 days stateside. The KC-135 flew to Guam and then the passengers were all bussed to the base ops terminal while the plane was refueled. I was the first to walk into the vacant waiting room and noticed a black nav bag sitting in the middle of the room. It looked oddly familiar. I checked it and found that it was my bag that had disappeared five months earlier. I checked with all the terminal workers and nobody knew how it got there. Nobody had handled the bag and it hadn't been there a half-hour earlier. David Copperfield would be proud of that trick.

As I remember, on the fourth tour our pilot Fred Green was replaced by Maj Tom O'Malley and TSgt Don Emerson replaced Joe Adams as Gunner. We were sitting on Guam waiting to see if the peace would hold. A lot of the B-52G aircraft and crews were then returned to the states to resume cold war alert but many of the B-52D planes and crews stayed. Upgrades and initial checkouts had been postponed during combat but now there was an opportunity to resume training. Some Copilots and Navs were sent over from the states to enter training since most of the D-models were on Guam. Our crew became the designated training program managers with the copilot and me in the upgrade program. As I was flying RN upgrade, we had new Navs flying with us including 1Lt Steve McCutcheon who joined the crew when I replaced Dick Kunz as RN. I remember I got real cocky on one training mission and decided to try to hit all five targets available at the RBS training site on Guam on one low level bomb run. It was a sync-alt-sync-sync-alt run. Managing five individual bombs in about five minutes is very hectic. It wasn't very pretty but all the bombs were within scoring criteria. Low level was a hoot too. We flew down the chain of Marianas Islands at 1,000 feet over the ocean and every 20 miles or so we used the Terrain Avoidance radar to pop up over the next volcanic island. We couldn't fly below 1,000 feet because of salt water corrosion hazards. I can't say that we ever went below 1,000 feet but the gunner says it throws up a huge rooster tail of spray.

We got home from that tour on 6 October, 1973, ready for our 28-day break. It was also a Jewish Holiday called Yom Kippur. That was when Egypt and Syria decided to attack Israel. All crews and planes were generated on alert for the crisis. The alert facility hadn't been used for almost two years and was filled to capacity. The VOQ was taken over as an alternate alert facility, complete with roving security guards. Because we had just deplaned, we were scheduled to take the last plane to be generated. There were so many planes generated for alert that there weren't enough alert parking spaces. Planes were parked in normally unused positions in the vicinity of the active runway. Maintenance had a lot of problems trying to generate our hangar queen. It had been torn apart for phase maintenance but they finally got it reassembled and ready for alert. It was towed to an open spot on the ramp and the nuclear weapons loaded. The plane was then fueled up to the full EWO (nuclear) fuel load before turning it over to the crew. One factor wasn't considered. That section of the ramp wasn't stressed for a 450,000 pound airplane and it sank through the asphalt. Because the plane was then almost sitting on the bomb bay doors, they couldn't be opened to remove the weapons. The plane had to be defueled, jacked up, weapons removed, the ramp repaired, the plane lowered off the jacks and towed to a more suitable parking position. Then the cocking procedure was started all over again. As I recall, the war was over before the plane was ready and we never officially assumed alert.

My final deployment to SEA ended in late spring of 1974. The powers that be had decided it was time to get the 7th Bomb Wing at Carswell back to its cold war mission. All the deferred check rides and training events were brought up current and we returned to back-to-back alerts. We were on alert when we had a visit by the SAC Combat Evaluation Group. We all tested well and flew an excellent mission. Our plane had an excellent bombing record so a lot of the betting pool was on us but we got edged out for the best bomb scores.

I recall an incident on one training mission that fall. It was a typical night mission with air refueling, low level bomb route, high level celestial nav leg and pattern work at home base. The north end of the runway at Carswell juts out a little into Lake Worth which is about a half mile wide at that point then the terrain rises quickly about 500 feet. We had flown several approaches to Runway 17 and still had another half-hour of pattern work left in the predawn hours. We were all tired from a long mission and I think I was getting a head start on my post flight paperwork while the pilots were on final to another approach. Suddenly the gunner called out to "PULL UP." He had just looked UP at the light on top of a water tower on the ridge north of the lake. I don't know how close we actually came to the ground but that approach was changed to a full stop landing and the mission terminated early. I was reminded that crew coordination doesn't ever end while you are flying.

I had been dating a wonderful girl in Fort Worth and at Christmas asked her to marry me. We were talking about a May wedding when I got a call on December 28 from SAC Personnel. It seems my records still said I was a SEA volunteer and they needed a volunteer for a short-notice remote tour assignment to Thailand in AC-130H Gunships. The Fire Control Officer position required a Bomb/Nav specialty code to direct airborne weapons. Although I had just completed over 700 days TDY to SEA it didn't count as either a remote tour or an oversea's tour. After discussing the situation with my fiancée, I decided to take the one year tour and get those blocks filled on my assignment record. It was only temporary. I would have a career broadening experience and get right back to SAC. Jana and I moved the wedding up to January 25th and reported two weeks later to the gunship training school at Hurlburt Field, Florida. After ten weeks of crew training, I deployed to Korat

RTAFB, Thailand. As I was deploying, gunships were involved in combat operations during the Mayaguez Incident. It ended up that there were no more combat missions for the gunships after that and many forces were being withdrawn from Thailand. After seven months, the base closed and the squadron was reassigned to Hurlburt Field. I was caught up in a unit move. Also, at the same time, the Air Force Chief of Staff decided that all future assignments would be based on weapon system worlds. I was directing gunfire from a C-130 type aircraft so I was classified for the next 18 years as a C-130 navigator even though I never navigated a C-130 in my life.

Alas, I had seen my last of B-52 crew duty and the glory days of SAC. Two thirds of the total mission ready flying time I had in B-52s was combat time. I lost track of most of the Crewdogs I had served with while I pursued a career in Special Operations. Actually, a career in the Air Force was the farthest thing from my mind when I enlisted. While flying in B-52s, my initial service commitment had expired but I decided to stick around until it wasn't fun anymore. After more than 24 years, the Air Force decided I had all the fun I needed and forced me to retire.

A B-52 Copilot's Tale
Jes Hales

This story starts as a B-52 copilot's tale and finishes as a B-52 instructor pilot's tale.

I graduated from pilot training in October of 1964 after spending 10 ½ months in class 65-XD at Laughlin AFB, Texas, where I flew the T-37 and T-33. My assignment after pilot training was to B-52Ds at Fairchild AFB near Spokane, Washington. Before that I had to complete survival at Stead AFB, Nevada, Nuclear Weapons School at Carswell AFB, Texas, and B-52 training at Castle AFB, California. I reported to Fairchild in March of 1965 with a pregnant wife and a German Shepherd puppy.

One of my first flights as a new copilot being checked out was with our squadron operations officer, Lt Col French. He was a great pilot and had been in B-52s for a number of years. What was so interesting to me about this flight was that it was a test of NORAD and the environment we found ourselves flying in to. We flew north to the Arctic Circle and then made a penetration run through Canada and into the USA. The flight was at high altitude and the aurora borealis display was something to behold. For a "good ole boy" from the South, it was very impressive. I had never seen the "Northern Lights" as they are sometimes called in the South. I can understand how a person can get disoriented in that environment. We had to trust our instruments because with the ever changing light show; it was not easy to know which way was up or down. To the best of my recollection, we were not intercepted by any fighters.

After completing the squadron checkout, I was assigned to a crew. All the other crew members were very well qualified in their positions and experienced. I was as green as could be and things did not go too well. The Aircrew Commander (AC) (AC-1) did not seem to have the time to give me the guidance necessary for me to be the copilot he wanted. There was a crew change and my new AC (AC-2) was very patient. We had many study session on the tech order - which helped me become a professional at my position. I'm grateful that he cared enough to help me become a good copilot and eventually a good instructor. The other members of the crew were also very helpful in getting me up to speed. As a copilot, flying the plane was no problem. Lt Col John Byrne at Castle was a great instructor. As well as being safe at takeoffs and landings, I was even able to hook up and refuel before I left Castle. Formation flying had been my best phase in pilot training.

AC-2 who had taken so much time with me decided to make a career change and resigned from the Air Force and went with the airlines. A new AC (AC-3) was assigned to the crew and in short order we found out we were to go to Guam as an Arc Light augmentee crew. It seems the cadre units did not have enough personnel to cover all the sorties that were required and other D-model units were tasked with sending crews to Guam to help cover the shortfall. We went over in January of 1967 and were there for 60 days and flew 16 missions. As the new guys, we always flew the number three position in the cells we were assigned. This was in the days before DASK and we flew some fingertip formation during the bomb runs to make sure the target box was covered.

Guam was a new experience for the crew and me. We were billeted in building 007, which was not air-conditioned at that time. One day spent there really comes to mind. I was trying to get some rest in the afternoon before a night flight when I woke up to find my bed moving around the room. I eventually came to realize we were having an earthquake. Before I could make my way to the door, it was over. My only other earthquake experience was at the Officer's Club at

Kadena in 1968 when the building started moving, light fixtures started swaying, and everyone headed for the exit door. Again, it was over as quickly at it started. Other times crews reported earthquakes at Guam while we were in the air on missions. I'm glad I did not have to experience those.

For recreation, we went to the beach, shopped at the BX, went to the Navy BX, toured the island, played golf, went to the O' Club for our meals and visited the bar on occasion. The officers' wives ran an outlet type business and we shopped there for clothes for our wives and other items to mail home. We also went to Joe's and Flo's for Mexican food and the hot salsa (equal parts jalapeños and tomatoes). Some of us even made it to the Navy O' Club for a meal.

AC-3 did not allow me to do much flying before we went to Guam on our augmentee tour and practically none at all while at Guam. The tour was not a pleasant experience for any of the members of the crew. When we returned to Fairchild, we found out the unit was scheduled to deploy to Guam in the Spring of 1968 as an Arc Light cadre unit for 189 days. I immediately visited the squadron commander and requested a crew change. I did not want to spend another tour with AC-3. The squadron commander said he would take it under consideration. Later I found out that each of my fellow crewmembers had individually gone to the squadron commander and requested a crew change. We got what we wanted; AC-3 was given another crew and we were given a new AC. The new pilot was John Carroll, an Air Force Academy graduate and one of the "sharpest" officers I ever met. John was a refreshing change for me and I got every other takeoff, landing, and refueling. I also got to fly the airplane at low level and on bomb runs.

We deployed to Guam in March of 1968 on a C-141. I was unfortunate enough to have the only airline type seat in the plane that would not recline. I did get a little sleep but very little. After getting settled in and going through SAC Contingency Aircrew Training (SCAT) school, we shopped for and bought a Guam BOMB - a 1957 Dodge. It ran well and the RN and Nav made sure it was roadworthy. We toured the island and enjoyed the freedom the car provided.

Missions at Guam were much the same as I remembered from the year before. On the first mission we had low oil pressure on one engine during climb out and John had me change out the gauge with one from the adjacent engine to try to determine if the gauge was faulty. It was not and the engine seized just as I had the new gauge plugged in. John looked at me and said we will have to abort the mission; I just shook my head - NO. One engine out was not cause for abort on those missions. He was reluctant to push up the throttles until the old Lt Colonel in the lead plane called back and said "You are going to have to over temp them sooner or later, might as well do it now and catch up." John was getting his Arc Light initiation. I never will forget the look on his face as the bombs left the aircraft. He was overjoyed to actually be in combat for the first time. Flying with John was a pleasure. He continued to let me do half of all the flying…takeoffs, refueling, bomb runs, approaches, and landings.

Several events come to mind about being on the crew with John. One night the refueling track had a lot of scattered thunderstorms and there was a lot of lightning. We were number three in the cell and arrived at the refueling track just prior to the tankers. The cell was in echelon and so were the tankers but we were at their 10 o'clock position and a couple of miles ahead. Instead of slowing and waiting for the tankers to pull ahead, the cell lead cleared the bombers to go to their tankers. Again, it was a dark and stormy night with a lot of lightning. We were afraid of being run over by number one and two because of the limited visibility. Following our Radar's directions, we finally got behind our tanker. We thought it was our tanker until we closed and a flash of lightning illuminated the plane(s) in front of us. You guessed it; it was number two tanker and number two bomber. John thought it was rather amusing and he got a good laugh out of it. We do not know what the number two gunner thought.

On another occasion, the crew decided to take our Guam Bomb on a tour around the island. At the south end of the island there is another small island about a quarter of a mile off shore. Some of the crew decided to rent a sail boat and go over to the small island. They never made it because they sank the sail boat at the dock. I have a picture of John and Jim the RN sitting in the sunken boat and the EWO also in the sunken boat laughing at them. So much of visiting the small island!

When we went to Japan on R&R we met Bob Ichamato (the gentleman who had decorated the Anderson Officer's Club). He invited our crew and one from March AFB to go out on the town that evening.

It was quite an experience because he took us to a real geisha house where we were served dinner and plenty of Sake and entertained. Bob's uncle, a vice president of Mitsubishi Steel, went with us. He was an elder gentleman and during the evening he sang a traditional Japanese Samurai song. After dinner, we attended a Japanese wedding party on the same grounds as the geisha house. Later on that same R&R, our crew drank all the beer in the Orion Brewery tasting room. Ugly Americans!

My good fortune was short lived. After we had been there about three months, the ops officer met us at the plane one night after a mission and yelled up the hatch, "Tell Hales there has been a crew change and he is going to Kadena with (AC-3) in the morning." I could hardly believe it. He had put me back on the crew with the AC I had flown with the year before. All good things must come to an end. The reason behind the crew change was because I was in the upgrade program and John was not an instructor but AC-3 was. I liked the new crewmembers and we got along fine except in our dealings with the AC. He was still hot tempered and difficult to please. Initially he did allow me to do some flying and I know he was impressed at how I had progressed. He would not let me refuel or make the bomb runs and in a short while he was again doing most of the flying. So much for being in the upgrade program and him being an instructor.

Toward the end of the tour, when my old crew and the new crew were both at Guam at the same time, I moved back in with John and crew E-15 because I was determined to get back home on time and the crews had different return dates. An official crew change was not made but the ops office knew what we were doing, and why, and did not interfere. When the unit returned to Fairchild, I was back on the crew with AC-3 that no one could get along with. Needless to say, I was not getting much upgrade training from him. When OER time came around, AC-3 had to go to his home of record on emergency leave. He had written all the evaluations on the crew just before he left. Two days after he left on emergency leave, the squadron commander called me in and said he had lost the OER AC-3 had written on me and he wanted to know what I had done the previous period because he would have to write my OER. I told him some of the things I had accomplished but that I could get my paperwork from home and give him more. He told me had had enough material. I do not know for certain but I suspect AC-3 gave me a bad evaluation and the squadron commander did not like it. It is rumored that when AC-3 got back off emergency leave, the squadron commander had him sign the new evaluation on me under the treat of giving him an evaluation like the one initially written on me if he did not sign it. I do not know if other members of the crew were treated the same way. Not long after this episode, AC-3 was give a new assignment and shipped off the base.

The crew was given a newly upgraded AC, Charlie Pierce. That sort of shot the crap out of me being in the upgrade program because Charlie was not an instructor either. Charlie was a great guy and let me have much of the flying but it was all from the copilot position. We were home during the winter and had to have CEG, ORI and Standboard flights during the snowy season. Fog at Fairchild was notorious and sometimes the runway would be clear at one end and be socked in at the other. We also had snow to contend with as well as alert to pull. It is amazing that everything was done before we had to deploy again to Guam. I enjoyed flying with Charlie and as I relayed earlier, he let me do much of the flying.

The unit again deployed to Guam in Mach of 1969. A couple of month into the tour, another crew change was made because I was still in the upgrade program and I was assigned as copilot on crew S-02. That put me on a standboard crew with an instructor, Neal Falk, and I did get a lot of flying and really enjoyed the new crew. A couple of missions with S-02 stand out in my mind. Flying out of UT on 20 June 1969 we had an unusual shaking of the plane as the weapons went out. After turning away from the target on our way back to UT, the comment was made that "that one did not feel right." The gunner chimed in and said "It was not right; two weapons came by the tail so close I could see the yellow rings around the nose of the weapons." The RN said the lights indicated we had a hanger. After getting down to 10,000 feet, the Nav went back to take a look in the bomb bay. When he got back on interphone his voice was a few octaves higher than normal as he reported we had one hanger and two riding piggyback on top of it and also one through the bulkhead. Neal made one of the smoothest landings I ever saw him make but the pucker factor was very high.

On another occasion the crew was selected to fly aircraft 5068 on a test hop at Guam on 17 July 1969 because another crew taking off on a mission had gotten into the overrun with that aircraft the day before. They had cut a tire and knocked down some ILS equipment. Our test hop was a great change from the norm. The aircraft performed well and we flew at low altitude up the chain to Saipan and Tinian and then back toward Guam. We were trying to burn off as much fuel as possible so we could land. To do this we went down to 100 feet AGL (water) and pushed the power up. We were doing 350 KIAS and the gunner reported we were pulling a rooster tail. The fuel was being burned faster than I could transfer it into the main tanks so we slowed down a bit. To make things more interesting, we made a practice bomb run on the ever-present Russian Trawler and opened the bomb doors. We never heard anything about it.

The new crew was selected to return to Fairchild about a mouth early to start preparing for the up coming ORI and CEG visit we knew would be coming after our return. Neal was great to fly with and I was making progress in my upgrading and passed my AC check ride. As it turned out, we flew the ORI before CEG. The Golden Wand system was used to clear the fog enough for the planes to launch on the ORI but we had to recover at Ellsworth AFB because of bad weather at Fairchild. Neal was selected for a school assignment but his departure date was not certain so another crew change was made. This new AC was Joe Dease, a former B-47 pilot who had come from of a B-52E unit before he was assigned to Fairchild. Joe was one of the smoothest pilots I have ever seen and taught me a great deal about flying the plane and some tricks about refueling that came in handy when I got my own crew and when I became an instructor.

CEG did arrive in the spring and the big lies were exchanged. They said "We are here to help you" and the unit commander said, "We are glad to see you." Our unit passed with flying colors. A major on the CEG team was to fly with our crew. We were told that he was really easy going and we should not worry. After the flight, we were told he was the hardest grader and most difficult of all the CEG pilot evaluators. The flight was on a Thursday before Good Friday. Our debriefing was to be on Friday. Early Friday morning (about 5:30 AM Pacific Standard Time) I received a call from my sister in North Carolina stating that our father had just died. I immediately called the chief of Standboard and told him. He said for me to call for airline tickets and that his wife would take us to the airport when we were ready. He also said my emergency leave orders would be ready for me to pick up on the way to the airport. We packed as quickly as possible and called and got scheduled out on a flight about 8:30AM. We were picked up by the Colonel's wife and driven to the airport and checked in. It was the beginning of a long day.

In the spring of 1970 there was an air traffic controller strike that started on Good Friday. We left Spokane on time and made a scheduled stop in Great Falls, Montana, but then had to sit on the ground for a long time. Finally we were airborne again but the pilot announced that he did not know where we would be landing - Chicago, Detroit, or maybe Cleveland. Our tickets were for us to make a connection in Chicago to take us to Raleigh Durham. After many circuits in the holding pattern, we were able to land at Chicago but we were so far behind schedule we thought we had missed our flight to Raleigh. We rushed as best we could with two small children to the Eastern counter to check on our flight. It was not on the board and we felt we had missed it. The attendant at the counter said not to worry, the flight had not been posted yet.

After a four-hour plus wait in the terminal, we were allowed to board our plane for Raleigh Durham. The plane was then towed to a parking area somewhere and plugged into a power source. We sat in the plane for another four hours before we were allowed to take off. We had first class seating because that was what was available when I called for tickets. I'm not sure if it was because of our first class accommodations or not but the attendants broke out the food and the booze. The first class section had a party for four hours. Betsy and I even recognized and talked to an acquaintance from our college days at East Carolina who was also on the plane. The flight to Raleigh Durham was incident free but we were many hours behind time and my wife's Dad had been waiting for us at the airport all that time meeting every plane and wondering what had happened to us. I do not know why, but no one told him why we were delayed. With cell phones technology of today, such a thing would not happen and he would have known what was happening with our flight.

After my Dad's funeral, we returned to Fairchild and I was assigned as Aircraft Commander on crew R-04. We were immediately scheduled to go to Guam as an Arc Light augmentee crew. That would be my fourth Arc Light TDY in a four year period. I liked my crew and we all got along well for the most part. My copilot was Steve Miller, the RN was Frank Claiaborne, the EWO was Eric Olson, the Nav was Nate Addleman, and the gunner was TSgt Potteroff. When we went through SCAT school at Guam a newly arrived Lt. Colonel was going to take the test with us but we answered the questions so fast he got lost in the dust. We had taken that test so many times we knew all the answers without reading the complete questions. The Lt. Colonel moved to another table and completed the test with another crew. The tour was for 90 days.

While at Guam, evaluation time on all my crewmembers came around and I had to write my first OERs as an aircraft commander. I had tried to prepare myself for that and I had copies of all the OERs that were written on me. I borrowed a typewriter from the admin section and started to work. In retrospect I must say it was sort of a crew effort. When I had completed the evaluations, I turned them in to the admin section. Immediately I was called down to the office to talk to one of the cadre ops officers. He told me in short order that I was giving my crew better ratings than they deserved and I would have to change them. I called my squadron commander at Fairchild and told him what was happening and he was not too happy that the cadre unit was trying to dictate evaluation scores for augmentee crews. I guess it hit the fan for a while but my evaluations were submitted as written and I do not believe any of them were changed.

During that tour I made the best B-52 landing I have ever made. We were at Kadena and I was flying with a Barksdale crew as a sub AC on a night mission. When I flared, I held the plane off the runway and was expecting the usual thump as the plane settled and as the rear gear touched down, I kept waiting and waiting and finally the Nav called up and asked if we were down. About that time, I felt a little rumbling as the gear rolled down the runway and I told the Nav, "Yes, we are down." I was shocked it was so smooth that we only felt the plane wheels rumbling as it rolled out. B-52 landings are not normally of the "roll on" type. The crew was impressed at the smoothness of the landing and I pretended that it was just a normal landing. I only wish I had made that landing with my own crew.

I was getting tired of going Arc Light every year and my family was not happy about it either. I wanted to transfer to a G or H unit and have some relief and family time. During the tour, a Lt. Colonel from Westover flew with us and mentioned that he was being assigned to SAC as chief of SAC assignments. The flight went well; the Lt. Colonel was impressed with our work and gave us a good write up. I mentioned to him that I was approaching 500 days Arc Light and surely would like to go to Seymour Johnson. He told me to call him when I had my 500 days. I counted the days and made the phone call. I got orders to report to the 68th Bomb Wing, 51st Bomb Squadron at Seymour Johnson in November of 1970.

After our Arc Light tour, but before my PCS, we were flying high runs on Wilder, Idaho, one night when I heard John Carroll on the radio. He had gone to Castle as an instructor the year before and was flying low releases with his student crew. I told him I was going to Seymour in November and he relayed that he was going to test pilot school at Edwards in the near future. I last saw John when he came through Seymour Johnson on a cross-country while in Test Pilot School. Upon his graduation, he went into the Steve Canyon program and was killed on 7 November 1972 after being shot down on the Plain of Jars. The enemy used his body as bait to try to lure the rescue choppers in. I do not think he was ever recovered. You can Google John L. Carroll and read about it. The Air Force suffered a great loss that day and I lost one of the best friends a person could ever have.

With today's interest in recent airliner crashes and the possibility of lighting strikes being the cause, I am reminded of a flight at Fairchild shortly before my transfer to Seymour. We were completing our mission and were shooting approaches at Fairchild in marginal weather. Winter had arrived in the Northwest and we were being vectored around snow squalls in the pattern and also using our radar system to remain clear. On downwind we appeared to be in the clear when we took a big lightning strike on the nose of the aircraft. I notified the command post and told them we would make a full stop on the next approach. On short final, we took two more strikes on the nose even though we appeared to be in the clear. After landing and taxiing to our parking area, Sergeant Bornstein was in the cockpit in a flash wanting to know what I had done to "his airplane." Upon inspection, the fiberglass at the ends of the wings and the top of the vertical fin looked like giant tooth brushes; it was shredded beyond belief. The lower nose radome had three holes where the strikes had hit. I guess we were lucky because we felt no ill effects inside the plane when we were struck. Those were the first and only lightning strikes I experienced in over 6,000 hours of flying time.

It was with mixed emotions that we left Fairchild. My wife, Betsy, and I really enjoyed Spokane and the beauty of the Northwest.

There were several restaurants in Spokane we liked to visit. The Ridpath Roof Restaurant had the best prime rib, the Longhorn had the best BBQ, and Hooligans and Hannigans had the best corned beef sandwiches. We had made trips to Seattle, Vancouver, Portland, and Banff, Canada, and found the cities and the scenery remarkable. Too, I really liked my crew and I knew I would miss them. At Fairchild I had grown from a very green copilot to be the senior aircraft commander on alert the week before I left.

The 68th Bomb Wing was much like the Fairchild unit except it was a tenant unit on TAC base. The 51st Bomb Squadron's claim to fame was that it had dropped more nuclear weapons than any other SAC unit. They dropped one near Savannah when they were a B-47 unit; they dropped two in Wayne County, NC, in 1961 when the wing came off a B-52; and they dropped four in Spain during a refueling accident. As a side note, one of the bombs in North Carolina landed in my father-in-law's bean field and he was the person who found it.

My wife and I both grew up in Wayne County, North Carolina, and that was one reason I wanted to be stationed there. We had been stationed 2,200 miles away when we were at Fairchild. Nothing disparaging about Fairchild -I loved it when I was there. The county is some of the most beautiful in the USA. I loved the mountains and lakes and really liked hunting and fishing. If you have not been to that part of the USA, you need to make a point to visit there. However, we really liked the idea of being stationed near home because we had been away from that part of the USA for the better part of six years. We bought a home off the base and we had a lot of catching up to do with family and friends. The children got to know their grandparents.

The G-model was a real pleasure to fly compared to the D. The upgraded engines made it more responsive and the lower pilot seat and larger front windscreen allowed me to see more of the tanker when refueling. Late in 1970 the Fairchild unit was transitioning to G-models and I went to G ground school at Fairchild before I transferred to Seymour Johnson. Major Paul Hamilton from Castle taught the G difference training. I later worked with Paul when I was stationed at Castle. My check out at Seymour went well and soon I would be on alert. Over the Christmas Holidays of 1970 an A/C died and I inherited his crew, E-11.

The unit expected an ORI at any time and it was not long in arriving. My crew was on alert at the squadron building when the klaxon went off. The squadron building was at the east end of the runway and the alert shack was at the west end. Jack Wiley was my copilot and he was driving the six-pack International truck we were assigned. We left the squadron building and started down the parallel taxiway toward the alert shack. Jack could not get the truck into high gear and we were just grinding away. I happened to be sitting in the middle in the front and I told Jack to push in the clutch. I grabbed the gearshift and pulled with all my might. We would either get it in gear or drop the transmission. Luck was on our side and it went into gear and off we went. Jack had the truck in a four-wheel drift going into the Christmas tree and slid up short of our plane, which was in the first stub. We fired up the plane, checked the systems and made our timing. Because it was a TAC base, I don't think we had to taxi. After testing, we went home, got crew rest and flew the mission the next day. I remember the target area because it was some big farm in Oklahoma. The farmer or someone had a large light like you see at grand openings and they kept pointing it in the direction of the IP. We easily could have bombed it manually if need be. There was a great party after everyone landed and all the good bomb scores came in. My crew had two good AGMs and four good bombs and received a blue ribbon for our efforts.

Being at Seymour Johnson and near family was great but there was one fly in the ointment.... my new crew was scheduled to go to Guam as an Arc Light augmentee in the summer of 1971. I was going Arc Light again! My crew was great and it was the first time they had gone Arc Light. Before going to Guam, the crew had to go through RTU at Castle AFB to get familiar with the D-model and the Arc Light mission requirements. We departed Seymour Johnson AFB on the 2nd of May, 1971 bound for Castle. RTU was like a vacation for me because I had six years in the D-model and 179 Arc Light missions.

We arrived on Guam on the 16th of May. The crewmembers were typical first timers when it came to shopping and exploring the island but we only had four days at Guam for SCAT School and then were sent to UT on a Young Tiger and started flying missions out of U-T on the 23rd of May. I started the tour with 55 wave lead missions to my credit because of my previous tours and most were when I was on a standboard crew. Soon we were leading cells as Wave Lead (WL) and

Airborne Commander (ABC). The missions were relatively short when compared to those out of Anderson. We flew several high threat missions, had many MSQ runs, had some lead changes due to bad FCS and radar systems and had our share of bag drags.

It was interesting that I, as a Captain, was leading a Major and Lt. Colonel on most missions. They both were former F-105 pilots. The Lt. Colonel did not like being number three most of the time and started giving me a ration of crap. In return, I wrote him up on the ABC report and talked to the briefing officers about his attitude and behavior. They told me not to worry, just keep doing my job, they already knew about the Lt. Colonel. Evidently his reputation preceded him to U-T. Having the two F-105 pilots in cell with me did come in handy one night on a high threat mission. We were to have two Wild Weasel F-105s and one EB-66 as support. About five minutes to release, the F-105s decided they were going RTB because each had some small system malfunction. The EB-66 said he was going feet wet. Both former F-105 pilots let the Wild Weasels know the problems they had were minor and to remain on station. As the ABC, I decided to press on because it was too late to call Red Crown or anyone else for clarification. Pucker factor was beginning to be high because the EWO said we were being looked at by the bad guy's radar. I'm not sure if the Weasels stayed or left. We did our job and got out of the area as fast as we could

On most missions crews never hear about the success or failure of their efforts, but we flew one on the 8th of July 1971 about which we did get feedback. My crew was Grape 1 and we gave the number two aircraft a bonus deal during the bomb run. There was a ground follow up and it was reported the strike had caused 37 KBA, destroyed 200 cases of mortar charges, 21 drums of fuel, 1,200 liters of oil, two 12.7 machine guns and three two-and-a-half ton trucks. I hope the ground forces benefited form the strike.

Jes receives 200-Mission patch from Gen. Frank Elliott.

On the 23 of July I celebrated my 200th Arc Light mission and was met by General Frank Elliott for the presentation of the patch and the usual wet down. General Elliott had been our Wing Commander at Fairchild and I really appreciated him taking the time to meet us and present the patch. Thinking back, I wish that I had requested a high-speed pass down the runway. I'm not sure they would have approved but it would have been a blast. It is reported that once at U-T, a crew was allowed to make a low, high speed pass and they had to pull up to clear the bomber turning off at the end of the runway. I guess that is why no high speed, low altitude passes were being flown.

We left U-T on the 11th of August 1971 on a KC-135 for Guam. At Guam we pulled a week of alert and then got ready to redeploy to the U.S. Our crew flew a bomber back to Westover AFB from Guam. We had two refueling and it took over 16 hours. My landing at Westover was not the greatest because I was really tired. The Westover unit was great. They met us at the plane, said for us to go to debriefing and they had a team to download all our "goodies" and gear from the bomb bay and elsewhere in the plane. When we got out of debriefing, the KC-135 to Seymour Johnson was waiting and all our gear had been loaded. We were on our way home and hoping that we had seen the last of Arc Light. We had seen the last of Arc Light but Bullet Shot was just around the corner.

On May 29, 1972 the 68th Bob Wing deployed to Guam as part of Bullet Shot. It was known as "The herd shot around the world."

Several B-52G units were deployed to Guam and the B-52 bombing in SEA had increased considerably. My crew was selected to fly aircraft number 571 to Guam. We were number two in a three ship cell led by a Standboard crew that, to my knowledge, had never been to Guam and had never flown a combat mission. I was hoping we could rotate the lead on the way over and give each crew an opportunity to lead. That was not to happen; the standboard crew led the entire flight.

To show my displeasure at them not sharing the lead, I showed them an Arc Light type rendezvous when we refueled. This type rendezvous was used by just about every aircraft commander who had spent any time flying Arc Light because pilots did not like to use extra time chasing a tanker down the refueling track. Time is precious when refueling. I left the power up when I descended, leveled off 1,000 feet below the tanker altitude at 330 KIAS, pulled the throttles to idle at one mile and coasted up into pre-contact position. I then went in and got my 75.000 lbs on the first refueling and on the second I did the same and took on 65,000 lbs. The RN on the third aircraft said he knew I was going to do it and when he saw me on the radar he said, "There goes Jes, doing an Arc Light rendezvous."

The flight over was about 16 hours and the crew made good use of the bunk area in the plane. Once when I went down stairs for relief, I saw the EWO had made himself comfortable on the floor behind the Nav and RN. I thought that was an unusual place to rest but he looked comfortable. About and hour out of Guam, the EWO came on the interphone and said he was sick and had fainted while downstairs. I responded that I thought he was sleeping there on the floor. The RN told him the same thing. I called the command post and requested a flight surgeon meet us at the plane when we landed. They wanted to know the nature of the problem and I just told them a crewmember was ill. The flight surgeon took the EWO when we landed and he never flew again and was soon sent back to Seymour Johnson. When we started flying, we flew with a substitute EWO for a while until our new EWO arrived from Seymour. He was Ken Voorhees and he was a great asset to the crew. As well as being a good EWO, he was a commercial pilot and gave us relief in the cockpit on the missions returning to Guam.

By the time Bullet Shot hit, I had 233 missions and many had been as WL and ABC. Because of my experience, my crew flew as WL and ABC most missions. We had our share of bad systems, lead changes, and bag drags during the tour. A couple of missions stand out in my mind. The first was when we had a triple bag drag (three airplanes). We were Lemon 1 and got the second plane to the number one position when the Nav called and said water was coming out of the spray bar and getting everything wet including his parachute and ejection seat. I advised Charlie and he said not to worry, the water separator would cut out at 25,000 feet. I ask him if he was telling me to climb to 25,000 feet in RAM with a soaking wet Nav in a soaked parachute and ejection seat. There was a lot of silence and he then directed me to taxi to a specific location and drag to aircraft 214. The other two aircraft took off on time and we were an hour behind when we got off the ground. The cell was given a 30-minute target extension and I had to make up 30 minutes. We got a short turn and raised the flaps on course. I found the G-model climbed really well at 310 KIAS and that is what I did. When we leveled off (I think we were at 34,000 feet), I allowed the speed build to 530 TAS. We did notice a few ripping and short jostling sensations that were not the norm. We finally decided that is what clear air turbulence (CAT) feels like at .92 mach in a B-52. Lead called back and said they were at base airspeed minus 50 KTS and wanted to know our speed. I responded that we were at base plus 80. We had 130 knot speed differential for the overtake. All I heard was a big "WOW" from the lead RN. We did make up the 30 minutes and were in position before lead called Bongo at coast in.

Another memorable flight occurred when a typhoon was threatening Guam. It was raining as hard as I have ever seen and heard. We were leading and at the number one position when I heard a drumming noise. I asked the RN if he was in sector scan because of the noise. He replied that he was not and what I heard was rain beating on the side of the plane. When cleared for take off, I aligned with the centerline lights, set the power, turned on the water injection and proceeded down the runway. The wipers were going as fast as possible. Soon the windscreen was gray with water and I had lost the centerline lights. I maintained heading by keeping the FCI centered, keeping my heading marker centered and watching the runway side lights in my peripheral vision to make sure I was not drifting left or right. This was a real instrument takeoff (ITO).

The weather was not much better when we returned and Charlie was broadcasting that it was 200 and a half. I flew an ILS with a radar monitor and had the RN put the cross hairs on the end of the runway. I directed the copilot to let me know when he saw the strobe light (rabbit). When he did call the rabbit, I came off the gauges and I could see the lights on the right side of the runway through the mist and rain. Since I was aligned with the right side of the runway, I corrected to get to center, overshot and put my left wing over the Charlie shack. I believe the wind had a lot to do with the overshoot. We were progressing toward the dip in the runway; I corrected with right rudder, got the plane over the runway and pulled the throttles to idle. The plane settled nicely and we got the chute out. I could make out the lights well enough to taxi. The two planes in cell behind me did not see the runway and were directed to a tanker to get fuel enough to orbit until the weather got better. I guess it was an eventful night because one plane is reported to have made a circling approach at the International airport. Circling at 200 feet with all those towers and hills around must have been sporting.

On a mission on the 3rd of December, Steve Miller, my old copilot from Fairchild, was flying with us. Steve was working in scheduling and went along to get his flight time. After landing, Steve was taxing the plane to the south ramp where it would be parked for refueling. The Sergeant who was marshalling us in evidently was not familiar with the turning radius of the B-52 and was bringing us in close and had not directed us to turn. I told Steve to go ahead and turn. When we were well into the turn I took the plane and stopped and signaled for the Sergeant to connect to the interphone. When he was on, I asked how much space we had between the wing tip and the blast fence. His response was "If you had turned when I motioned for you to…." At that time Steve said, "B--- S---, you never told us to turn." I opened the window, requested the hatch be opened and I went downs stairs and out to the end of the wing to see for myself how much space I had. It was less than four feet. When I got back in the pilot seat, I used the crosswind crab system to move the plane away from the fence and into proper parking position. As I was moving, I heard the pilot in the number two plane tell Charlie that he was going to shut it down on the taxiway if that Sergeant was to marshal him in. They got someone else to marshal the other planes. After debriefing, we talked to the duty DM and tried to explain what happened and how a B-52 should be turned to maintain proper position on the taxi lines. Evidently he did not listen to us because two nights later, a B-52D was put into the blast fence at the south ramp. I'm sure some poor pilot got blamed for it.

On the 8th of December I was sent to CFIC at Castle AFB in California. I rode a KC-135 to Travis AFB and then a bus with several Air Police and an attack dog to Castle. I really enjoyed CFIC and learned a lot about flying the plane and I got to see many of my old friends who were stationed there. Because I was at CFIC, I missed the December bombing of North Vietnam. I did follow it in the news and was concerned for my old crew and my friends. I returned to Seymour Johnson on the 21st of December. During the holidays, I saw a captured airman on the news that I recognized as a Seymour Johnson gunner. He had flown as a sub and the crew had been shot down. I reported this to the unit and found that several other people including the gunner's family had also seen him.

I worked as assistant to the ops officer at Seymour Johnson and did not get back to Guam until the 13th of April 1973. I took an instructor crew and we flew a lot of training missions to get the new crewmembers upgraded. We would fly a combat mission about once a week and take the new people upgrading to let them see first hand what happened on a real mission. I remember one training mission when we demonstrated a minimum interval take off (MITO) for the upgrading pilots. I was number two and was 15 seconds behind number one when I crossed the hold line. When the planes went by the tower, I was ten seconds behind lead and number three was eleven seconds behind me. They talked about that MITO at Andersen for some time. We had three BUFFs on the runway below S-1 going like bats out of hell.

My wife and children came to Guam for a month and we rented an apartment on Tumon Bay and I bought a '74 Chevy wagon. It was pretty nice as Guam Bombs went. It never let me down. The family really enjoyed the beach, swimming, going to the O'Club and shopping. I soon received orders that I would be going to Castle AFB as an instructor in August. Our crew was coming up on leave and we were torn between staying on Guam during leave or going back to Seymour Johnson and getting ready to move. Finally, we decided it would be best if we went home and got ready for our PCS. We sold our Guam Bomb and gave up our apartment. The family went home on a commercial flight and I went on a KC-135. Our scheduled customs stop was at Carswell AFB, Texas, in the middle of the night. A Sergeant on the flight had bought his dad a bottle of Royal Salute Scotch and it was in a nice velvet bag. Somehow the crockery bottle got cracked on the shoulder and upon inspection it was decided that no shards had gotten in the scotch. We had a party on the ramp in the middle of the night at Carswell drinking scotch out of a crack on the shoulder of a crockery bottle. It was really smooth.

I arrived at Seymour Johnson on the 26th of June and the family decided we should go to the mountains of North Carolina for a mini-vacation. While there, my sister called and said a Major at the base wanted me to call him. I was not too happy because I wanted my leave and vacation and did not want to go immediately back to Guam. When I talked to him, he said I would have to come back immediately. I replied that he was going to have to tell me something other than that to get me to come back. His response was "We tested the Klaxon today." My response was, I'll be home tomorrow." I knew the unit was coming home and I would be the first crew to fly a training mission and go on alert. One problem was that my RN had stayed on Guam for his leave and no one could find him for a while. He eventually made it back in time to go on alert.

I finished Arc Light/Bullet Shot with seven tours, 300 missions, 124 ABC sorties, 108 Wave leads, and 864 days TDY between 1967 and 1973. I have over 4,300 hours in all models of the B-52 except the A and E.

I was sent to Castle AFB, California, in August 1973 as an instructor and remained there for two years. While there, I made Major and was selected to attend Air Command and Staff College (ACSC) in the class of 1976. After my student year at ACSC, I remained as a faculty instructor (seminar leader) for two years and then was in charge of the Big Stick War Game and my final year I was in charge of Strategic Forces Curriculum. That year the Strategic Forces Curriculum and Strategy and Doctrine were dovetailed and received the highest rating of any phases of instruction in ACSC history. My phases of the curriculum were tied heavily to my experience as a SAC crewmember but that was to be the end of my association with SAC.

My follow on assignment until I retired in 1984 was to MAC as an instructor pilot in the T-39 unit at Maxwell AFB ferrying General officers and other dignitaries around the country.

Rated Supplement
Pete Seberger

In October of 1978 I finished a short tour at Clark AB, and since SAC never lets go I was notified to expect reassignment back to the SAC crew force. At the time the Vice Commander of SAC was known to have made the public statement that the place for passed-over officers was on alert as role models to the younger crew force, so I selected Ellsworth AFB, as they flew H-models and most of my time to date was in that model. It was also the closest base to my home in Nebraska, and my parents were getting on in years so I wanted to be closer. I was unfortunate enough to have been at Castle when the new unit quota 1-2-3 OER system, which gave birth to the largely fictional "Promotable Three", was instituted, and had failed promotion twice during my subsequent short tour. I did not object to returning to the crew force, since I never saw a staff job in SAC I thought I would prefer to flying, and I considered being an instructor in the BUFF a good job.

When I arrived at Ellsworth, it was two months after school had started for my children who had remained at Merced while I was in the Philippines. My son had been placed in an advanced class in the fourth grade for something called MGM, short for mentally gifted minors, in which he was blossoming. My wife and I decided to let him finish that year before moving the family to Rapid City. I found an older motel that had been converted into apartments in west Rapid City and spent much of my free time looking at houses. I also began requalifying in the H-model, and since I had only been out of the plane for 15 months and change, SAC allowed me to requalify simultaneously as AC and IP. As the IP and Stanboard shops were already full with local boy-wonders and I was naturally on the bottom of the pecking order, thanks to the aforementioned command attitude, I expected to go on a crew, which I did shortly after requalification. I was no stranger to alert, but it had been nearly seven years since I had pulled it regularly. My last 18 months at Grand Forks as an IPU, the four years at Castle, and my Philippine tour gave me a nice long vacation. And, since my family was not yet at Rapid City, alert was not as burdensome as it was once.

When I eventually went on alert one of the first sorties I certified was, as I recall, called an "R and R". No, not rest and recuperation - it meant recovery and reconstitution. This was a sortie for which directives specified the crew perform evaluation of the target prior to release to determine if the target had already been hit, and if so to withhold that weapon for another target. As I recall there were several more targets than weapons, but the Higher Headquarters expectation was still that there would be weapons left unexpended. After landing at the post strike base the crew was supposed to contact or be contacted by whatever NCA survived, and then executed on a restrike sortie from that base if weapons, plane, fuel, and crew were capable. To that end, additional worldwide maps and targeting materials were carried on the plane, stored in several aluminum boxes similar to those which contained the CMF materials in the plane for the initial mission.

Those of you who ever pulled alert know that after loading the mandatory alert gear, the CMF, and other professional gear, there is not a whole lot of extra room up front in an H-model. There certainly was not room for those extra boxes of maps, so somebody (never found out whose idea it was) had the bright idea that they could be carried in the 47 section. They would be out of the crew's way during normal alert and accessed from the ground after landing if they were needed. As I recall, each box weighed about 40 pounds and there were six or seven of them. The only place they could be conveniently placed was on the skin of the airplane in the vicinity of the oxygen converters. So, the first morning I pulled alert it was my duty to inspect those boxes along with the normal daily alert preflight. That inspection was considered necessary only once and I suspect that many crews did not bother to inspect it at all, other than perhaps a cursory look. The contents of the boxes were never changed, inventoried, or signed for by the crew on alert. I think someone in the bomb-nav shop did the deed, I just don't remember if we were ever told.

What alarmed me that day was the fact that on this particular airplane the aluminum boxes with sharp corners and considerable weight were tied to the oxygen converter coils through the handles with quarter-inch manila rope. This was supposed to keep them from moving around when the plane was moving. At the time (early 1979) there were 30 planes in the two bomb squadrons at Ellsworth, and I think there were about half on alert at any one time, so a COCO exercise could get right sporty, to say the least. I know that despite my smooth taxi skills I was afraid that it still might be possible for some damage to occur near those 75 liters of LOX, and given the oil for cooling the ECM equipment and other flammables in that end of the plane I thought there had to be a better way to secure those boxes. So I immediately began an attempt to fix things.

I knew that drilling holes or modifying the plane in any way would require manufacturer interaction (and would take years) so I drew up plans for some homemade racks which I thought would work well. My plan was for two long pieces of relatively heavy aluminum angle to be clamped on each end to the main longerons which spanned that section about two or three feet above the oxygen converters. The clamps could be shop-made from aluminum and would not require drilling into or in any way modifying the longeron. Then the boxes could be secured with either a net or additional clamps to the aluminum angle, and they would not represent a threat to any of the plane's equipment. They were nice plans, and I thought it would be easy for the maintenance shops to build, as there were only two sorties on alert which would require them. So much for theory.

I contacted the various shops on base, talking with the senior NCO of each, and later with their superiors. I ran into a stone wall in each instance, culminating with the bomb squadron commander and ADO, each totally uninterested in "rocking the boat". Of course neither of them was pulling alert. I considered going higher, and figured that even if I was successful there would be too much face lost by those I had already talked to, and that they could and would make my personal and professional life hell for the years I had left till retirement. After that, I started looking for a way off alert. I was sure that it was only a matter of time until somebody tore something loose and started a fire or worse. That was one time I am glad I was wrong.

Sometime later that spring of 1979 I read an announcement that the position of Physiological Training Officer was open at Ellsworth, and rated officers were encouraged to apply for it. I checked out the position, and thought it presented some opportunity - not for promotion but a way out of regular crew duty. However I still wanted to fly so I had to find out if that was possible in that position. At the time, the Ellsworth Wing Commander was Al Renshaw. He and I had been roommates and on the same student crew in 1964 when I was a newbie at Castle and he was a B-47 copilot upgrading to the B-52. There had not been an AC on our student crew, so both of us got some time refueling and our share of pilot work. We had gotten along well, surviving a near miss with a traffic barricade in his MGB one foggy morning at Castle.

I applied for the PTO position, and was told that SAC would not release any IPs to the rated supplement as they were needed to fly. I got an interview with Col. Renshaw and asked him if he would query SAC as to the possibility of me remaining on attached status as an instructor. It had never been done for that position, but I was willing and thought that even (especially!) if I could not pull alert I could still do some instructing if I was freed of normal crew duties. SAC's reply was that they would not allot any extra flying time but if the Wing could provide the flying time for my currency and evaluation out of their present allotment SAC would agree to release me from the crew force and to fly as an attached instructor. I was a bit surprised, as the aforementioned SAC Vice-CC had well known views for us "Passovers." Perhaps it never crossed his desk. Anyway, I was cleared for the assignment and given one week (on alert, of course) to write OERs for myself, my crew, and each of my four flight's ACs, along with two suggested endorsements for each. Talk about creative writing! I had never flown with any of my flight's crews, and had only 90 days supervision, but I got them all written during that one tour.

In June of 1979 I got off alert one morning and that afternoon took a flight to Brooks AFB in Texas, where the school for Physiological Training Officers (PTO) was conducted as an annual class. I was one of about 15 students and we had a pretty good time practicing classroom instruction and learning the material. Following the school there was an additional two-week course for PTOs whose unit also had a hyperbaric pressure chamber, so I stayed for that course as well. When I returned to Ellsworth I got really busy. My family was still in Merced and when I got the PTO position we delayed their move until late summer. We moved into an apartment in the section of town we planned to purchase a home to give us time to choose. It was January of 1980 before we found and purchased the home where we now live. It was not a good plan as it turned out, since interest rates were 11 percent when we finally bought. It was 1992 before I refinanced, down to eight percent. Ahh - the Carter administration!

I found the work running the Physiological Training Unit to be challenging at times, but enjoyed having the time free from alert and some control over when I flew. There were always two PTO officers qualified to teach or monitor the teaching of the classroom courses, monitor the chamber flights, and operate the hyperbaric chamber for those medical emergencies for which the chamber was available. Our workload at the unit did not require a classroom presence every day, and we operated the chamber as necessary, but again, not every day. Accordingly, my NCOs pretty much ran the maintenance and operation of the enlisted functions and myself and the other PTO shared supervisory and classroom duties. The other PTO was nonrated so I could schedule him to do those required functions whenever I was scheduled to fly. I had a good relationship with the schedulers, so I got even more flying than I would have if I were on a crew. I perceived that my status as an IP and command pilot also gave me more credibility during refresher courses. I preferred to teach the one day refresher courses, since our courses for original training did not include regular flight crew but rather crew chiefs and others on flying status or scheduled for a passenger flight. I retained my line badge with escort authority to the alert facility, so I even scheduled classroom training at the alert facility for refresher courses sometimes. That let the crew member come in for just the chamber flight later.

Another policy I had was to take newer enlisted men assigned to the PTU out to the alert facility for lunch sometime during their first couple of months. Assignment to the PTU was a hospital position. My people wore whites, and I thought people used to a normal work day and week in air conditioned offices might better appreciate the long hours and sometimes harsh conditions the enlisted OMS troops had to endure, not to mention the gate and airplane guards. After those visits to the alert facility the new guys invariably spread the word that there were others worse off than themselves. At the time we also had an ACE flight operation, in which tactical airplane copilots flew T-38s on flights to build flying time as PIC. I asked and got permission for each of my enlisted troops to fly on each of the airplanes flown at Ellsworth; tankers, B-52, helicopters, and the T-38. The RC-135s airborne command post and recce mission did not permit passengers. My reasoning was that they were all trained to give some classroom instruction to new students going through the chamber to get on flight status and they should have that experience themselves. I insisted that if they flew any airplane they had to fly them all, and virtually everybody wanted a ride in the T-38, which the attached instructors were only too happy to provide. Some of my troops balked at riding in the B-52 since it was usually a long flight, but eventually went and I think they gained an appreciation of the long hours we Crewdogs put in.

One particular memory is of coming in one morning about 11 after landing at oh-dark-thirty earlier from a night mission. One of my junior troops watched me come into my office and said loudly enough for me to hear "Wish I could get officer's hours." I called him in and asked him if he would like to have the same hours as mine the next week, and he agreed. I told his NCO supervisor to make arrangements for him to have my duty hours. Then I made my flight schedule for the next week, which included an early morning show for a long flight. Nothing more was said till the afternoon before the flight. About 1300 I told the airman to come along with me, we were going over to the squadron mission planning room to brief the next day's flight with the flight crew. About 1500 we were finished. I told the airman we were done for the day and he was free to go home or whatever, but he could do no drinking and was to report to Base Ops at 0330 the next morning. He blanched and complained that he had no transportation since he car pooled with somebody else. I arranged to pick him up at his apartment at 0300.

We showed at Base Ops at 0330 for a normal weather briefing etc. for a 0600 launch, followed by an 11-hour flight. He was bored stiff after an hour on the plane, but stayed out of the way, and when we landed he had arranged for a buddy to pick him up. I told him he wasn't free yet; we had maintenance debriefing, followed by a crew critique which I generally had with the crew - especially the pilots. That lasted an hour because there were two or three pilots who had some questions for me and I had some for them. Long story short, he was free when I was - about 1900. As I dropped him off at home he said he would just take the next day off because his NCO's policy was to give time off if they had to work overtime. I told him to be at work at 0700, as I would be since I was teaching a refresher course that day. He blanched again, as he had evidently spent most of his airplane time planning how he would use his day off. But he was there when I showed the next morning and I made sure he stayed all day and went into the chamber, as I did. Not surprisingly, when the word got around, for the next three years I never heard another crack from anybody about my hours.

Another policy I started was to schedule weekend refresher courses for Guard or Reserve units who generally could only get airplanes for transportation to the PTU on weekends. My NCOs were resistant to that because their enlisted men sometimes had weekend jobs. I told them to schedule themselves, and the men who manned the chamber on a weekend, for a full day off within the next week, and to take volunteers. I would teach the refresher courses myself in the morning with a single enlisted clerk to handle the paperwork so each would only work about three hours. As soon as that became policy, there were more volunteers than they could handle and there was even competition among the troops to get in on the good deal. Often the units would cancel a couple of days out because of the lack of an airplane, and when that happened the guys who had volunteered were disappointed. I reminded them of the basic SAC Maxim, "You can't get screwed out of something you never had." But we got great feedback from the units who scheduled, even if they had to cancel, as we were the only unit who would even consider it. By the time I retired that policy had to change somewhat since the Flight Surgeon's office was required to provide a doctor to be in the office during chamber flights. That was fine during the regular work week but they balked at having to be around Saturday afternoons.

The PTU at Ellsworth also had a hyperbaric chamber, a result of the fact that early in the operation of the B-36 and B-52 it was fairly common for pressurization to fail and flight crew to have problems which could be treated in hyperbaric chambers. Some of those were actually a result of altitude chamber training. The Ellsworth runway was also longer than most, and Ellsworth was an emergency landing runway for the Space Shuttle, which began operation about that time. My PTU troops were trained to operate the hyperbaric chamber and we conducted training operations weekly including dives which used standard Navy decompression tables. There were/are some fairly exotic disorders for which the chamber is useful, and we maintained the capability to assemble a team for operation of the chamber with fairly short notice. That caused some difficulties at times, since it was well before the time of cellular phones and the officers only had a single pager, carried by whoever had the duty, so to speak. We depended on the fact that a dive crew was four people and there were about 16 assigned to the unit who could fill any position to be able to contact enough people in a short time. In four years there were only about three incidences when we had to assemble a team, and always had sufficient notice to be ready before a patient arrived. Our practice dives were rather fun, as we usually had at least one dive to 165 feet and nitrogen narcosis at that depth produces a high similar to three or four beers, with no hangover. There were lots of volunteers for those dives.

Another facet of that PTO position was that the PTU at Offutt was considered a satellite of Ellsworth. While there were some permanent maintenance people at Offutt assigned to the unit, all the operations and classes were conducted on a TDY basis by the Ellsworth PTU personnel. My parents lived in Nebraska, roughly midway between Offutt and Ellsworth, so I generally taught the three-day original course and two one-day refresher courses at Offutt, where we manned the chamber one week each month. It was convenient for me to leave Rapid City the Saturday afternoon before a Monday class, stay the night with my folks, then drive on Sunday to Offutt and get a good night's sleep before starting on Monday. My enlisted troops did not have to be there for a chamber flight till Tuesday, so they generally drove themselves by POV for the convenience of the government on Monday, and split costs. That left us local transportation in our own vehicles, not paid for by the government but convenient for us.

The Ellsworth transportation squadron was not able to provide a government vehicle for our travel, let alone two of them, as they were busy with support of the 150 Minuteman 2 missile sites and many launch control facilities around Ellsworth. POV for the convenience of the government saved 75% of total travel costs, and the hospital commander, my boss in the chain of command, was only too glad to have those travel funds for his doctors and nurses to use for continuing education travel. That was the routine until around June of 1983, when an NCO at the travel section of the finance office at Ellsworth reviewed my travel itinerary and noted that I was visiting my home of record each trip. He was convinced that I must be "screwing the government" because of that and told me so. I could not and did not ever claim anything for extra travel mileage or overnight lodging at my home of record. I showed him how much money the government saved, compared with the normal procedure of using civilian airline transportation (which we were required to budget for). It didn't matter to him, though, he still notified the IG, who conducted an investigation without ever talking with anybody in my unit. They forwarded their findings to my boss, the Hospital Commander, and to the Ellsworth Base Commander, a former SR-71 SRO as I recall, who told me to my face that I must be screwing the government and to quit doing what we were doing. Accordingly, I notified my boss that since I had a letter from the transportation squadron refusing my request for government vehicles we would all be flying from then on and to plan accordingly. It would cost us an additional two travel days each trip due to the restrictions on flying after chamber flights and the poor weekend connections to Omaha by commercial plane from Rapid City, not to mention the local taxi and other transportation we would be claiming. He got pretty upset, not on our account but for his doctors and nurses, and finally got the transportation squadron to give us one vehicle for my enlisted troops to use. Because of the complex regulations for using government vehicles, several breakdowns in the first couple of tries and other problems, our transportation to and from Offutt got very complex and expensive. We had to cancel at least two chamber flights for lack of personnel due to government vehicle or commercial plane troubles, and I retired before a permanent resolution to the problem was found. So much for fixing what ain't broke.

The biggest disadvantage to the PTO position was/is the chain of command. It was/is considered a hospital position, and the hospital commander becomes the reporting official and his boss the next and so on. At Ellsworth, that left me serving the 28th bomb Wing as an IP for nominally a third to half my time, but no input at all from them to my OER. My boss for the first three years was a 300 pound chain-smoking doctor (not a flight surgeon) who, to my knowledge, never set foot in our facility 80 yards from his office. His boss was the aforementioned Base Commander, and his boss was the Missile Wing Commander, who had a policy that no passed over officer's OER ever got forwarded past his desk. So from a promotion/career standpoint it was not a good move and I would not recommend it to anybody seeking promotion. In my case, that was no longer a consideration given prevailing attitudes at the time.

During my tenure at the PTU I prohibited smoking inside except for the break room and certain offices. That was an unpopular move, and perhaps contributed to my boss's absence. However, around the time I retired I think SAC began the policy of no smoking on SAC aircraft, and fortunately the trend continues. In my flying career in the BUFF, I had four catalytic filter failures, and smelled each one in time to shut off the air conditioner before the filter failed completely. I always thought then that smoking on the airplane was a bad idea, but it was, as all you old Crewdogs know, prevalent way back when. When I was a copilot my RN smoked the most wretched cigars, and the navigator chain-smoked cigarettes. Fortunately the AC did not smoke. I think I used 100% oxygen more than half my time on the plane. Last time I visited an airplane, at Barksdale in 2002, I thought the plane was cleaner then than when we flew them. I generally found an excuse to depressurize the plane and open the pilot or copilot's sliding window at least once on downwind on each flight in order to suck out the dust and crap that accumulated around the cockpit. Once I even threw out a paperback that the extra copilot in the IP seat was reading while we were on downwind. I told him I wanted six eyes outside if I could get it and always, on my own and student crews, expected the EW to read every checklist he was permitted to read to free up extra pilot eyes to spot other traffic or perhaps catch mistakes.

One thing I had to do to maintain qualification in the plane during my tenure as a combination PTO and IP was take the usual routine of Stan-Eval rides. Unfortunately, I never got a flight to myself, but generally tagged along with another crew and evaluator to complete some portion of my checkride so I might be evaluated during parts of three or four different flights. In January of 1983 the 28th bomb Wing began sending some of their planes to other units in preparation for the eventual assignment in 1985 of the B-1 and they had a sufficient number of IPs available during that transition. It was also time for part of my annual evaluation so I was one of three pilots during an enroute descent into Ellsworth and we did a double seat swap where I got out of the right seat and another pilot got into the right seat, after which I was supposed to get into the left seat as I was supervising the right seat pilot's approach and landing before doing one of my own. With four pilots aboard (including the evaluator) that got a bit crowded. During the transfer we were descending but still above 25,000 feet, so I was supposed to plug in my helmet to an oxygen line while the left seat pilot was unhooking. The young pilot evaluator managed to find that I did not do that within the timely manner he thought appropriate, so he got the thrill and bragging rights of flunking the old IP for an oxygen regulation violation. As my position as the PTO unit commander was well known, that violation had a delicious irony. Of course it did not have to be corrected by flying again, but only required ground corrective action which was a good lecture by another IP on the need to observe oxygen regulations.

I am sure the (comparatively) young evaluator was commended for his diligence, and I would like to believe that he had not been instructed by the wing to find some reason to fail me so I would go off flying status voluntarily, but that was the result. Hell of a note, I thought, no FINI flight, no wetdown, no final landing, no picture to show my grandkids, what a way to end a flying career. One of my contemporaries, the base instrument school instructor who was also a high time BUFF IP and a passed over Major, (now dead,) was appointed to perform the corrective action. He read the evaluation and recommended corrective action, scowled in disgust, and signed it without comment. I went to the appropriate offices, turned in my line badge, and went on excused flying status.

One relevant thing happened when I was at Ellsworth; I think in early 1979. Victor Belenko, the MIG-25 pilot who defected into Japan sometime in '77 or '78 was making a tour of US bases, and at the time his visits were classified. When word was passed for a general crew meeting it was somewhat of a surprise when he came out to talk. One thing he said in particular was that life in the Russian Air Force was like being a hen in a pen full of roosters. You knew what was going to happen, just not when or by whom. I always had the philosophy that evaluations were like that, and the game was to make sure that when you flunked or were downgraded it was for a reason and that both you and the evaluator knew. In my career, the most reasonable evaluator I ever had was one day at Castle when my copilot (also an IP) and I were getting a simultaneous evaluation, from the Castle Stan-Eval section. We pulled up behind the Castle tanker, and the tanker IP told us he had five student boom operators on their first flight aboard, and requested max contacts. The evaluator leaned forward and said to both of us "You guys both get highly qualified (H) in air refueling, you do whatever you want. Perry (the other IP) and I looked at each other, took the evaluator at his word, and each of us over the next hour and ten minutes on a double track made 15 to 20 contacts, demonstrated limits to each student boomer, got the tanker student pilot and copilot some autopilot off time and otherwise had fun. The next day the tanker pilot looked me up and said that was the best student training sortie he had ever had. And we did both get H in air refueling, something else was found to keep us out of H overall. After 20 years of evaluations I observed a universal rule. Evaluators keep a sharp eye till they have reason to keep you from being rated H overall, but try not to see things that would flunk you unless it's really unsafe or they have a quota to fill. After all, they might have to pull your alert tour while you are requalifying - cluck, cluck, cluck.

The Saga of Tail End Charlie –
(Not Your Normal Crewdog)
Geoff Engels

How many Crewdogs do you know of who have been fighter pilots, Forward Air Controllers, and Electronic Warfare Pilots, as well as having flown the B-47 and two models of the B-52? Know anyone who was awarded the Distinguished Flying Cross in both the smallest and largest combat aircraft used in Vietnam as well as the Silver Star and flew over 600 combat missions? How the heck did something like this happen?

Well, it wasn't my idea. I didn't start out to be a Crewdog; in fact, I never wanted to be a Crewdog. I always wanted to be a fighter pilot; however, it seems that the "Powers That Be" had other ideas. It all started this way:

When we were in school (USAFA, Class of '62) a SAC General came to give our class the SAC speech. When he was finished, he asked if there were any questions. There were. However, they were not the ones he expected. They were sharp, sometimes critical, and some were hard to answer. Since this was normal for our class, who were nicknamed "The Red Tag Bastards" (referring to our class color and nametags), we did not think much about it. The General apparently did. He actually liked the fact that we thought for ourselves and decided that our class should go to SAC. And so it came to be that when we graduated from pilot training, 85% of the class went to SAC.

Presented with this turn of events, my buddy and I retired to the O'Club for some thought and refreshment. Having had sufficient refreshment, we analyzed our options. We were high enough in the class so we could choose from the KC-97, B-47, KC-135, and the B-52. After some more refreshment, we looked at it this way. Since we both wanted to be fighter pilots, we knew that if we chose the B-52 or KC-135, we would be in SAC for good. However, the KC-97 and B-47 were known to be headed to the boneyard. We reasoned that when they went, there was a chance we could get out of SAC. Since the KC-97 was prop and the B-47 jet, we opted for the B-47.

And so it came to be that we became copilots on the B-47 at Pease AFB, New Hampshire. The aircraft was so near retirement that there was no longer a Combat Crew Training School (CCTS) and our entire checkout and training was OJT. We all knew that there was no intent that we ever upgrade to Aircraft Commander. After about 2 ½ years, and a few interesting experiences such as having the canopy depart at 33,000 feet when I had just finished taking a sextant shot and was therefore not strapped in, and performing a number of ATO assisted MITO takeoffs, before the retirement finally came. The canopy loss remains unexplained. We even tried to reproduce it in the simulator, but were unsuccessful. However, I did get my own safety supplement to the Dash-1. Something about installing seat pins before taking sextant shots, I believe. The ATO MITOs were very interesting. For those of you who never flew the B-47, it could be equipped with a belly rack that had 30 rocket bottles of 1,000 lb. thrust each for 15 seconds for an assisted takeoff. Needless to say, when you lit them off, you were committed. The MITO meant that the runway spacing between aircraft was reduced to 7 ½ seconds. We believe they set the takeoff order by averaging the number of children the crewmembers had. The more kids, the closer you were to the front of the stream. Since my crew had two bachelors, we were "Tail End Charlie" for all the launches. It seems that SAC had tested ATO takeoffs with the full 30-bottle rocket rack, but had never tested them in MITOs. Since this was the way an Emergency War Order launch was supposed to occur, they picked us to be the guinea pigs.

Interestingly, the daytime launches were easier than a standard MITO. The six J-47s with water injection made a lot of smoke and if you were at the back of the stream, visibility sucked. The rocket blast did not make much smoke (smokeless powder?) and was angled so that it actually blew the smoke to the side, making visibility better than normal. Nighttime launches were spectacular. The exhaust was almost blindingly bright. Once I made the mistake of looking back at my own rockets. It looked for all the world like the entire back of the aircraft had blown off and was burning joyously. After that I stopped looking. The last nighttime test was different. Weather conditions were such that ATO fog developed. For those who have never heard of it, ATO fog occurs when the rocket exhaust condenses instantly into a very thick cloud (similar to a contrail, but on the ground). From the side it appears as if a line of thunderstorms builds up on the runway. Since there were six or seven aircraft ahead of us, all adding to the density of the fog, when we entered the cloudbank and fired our rockets, visibility went to absolute zero. From the time we hit 70 KTS to the time we were 200 feet in the air we could see NOTHING! At the post flight debriefing, one of the ACs opined that he thought that if he had turned out his landing lights, visibility might have improved, but he didn't want to do it again to find out. Amen to that!

When the B-47 retirement was formally announced, once again we retired to the O'Club for more thought and refreshment. That time we reasoned that we probably could not get a fighter; however, since we had flown them in pilot training, we might get an IP slot flying T-38s. Thinking that this would be the best stepping-stone to fighters, we applied for ATC.

Eventually the long awaited assignments finally came in. That brought an instant trip back to the O'Club as we both got F-100s. Party time!!! Dumbfounded and hung over we finally found out the reason. There was a rule in effect that said that no one could be sent back to Vietnam for a second tour until everyone else had their first. They were flat out of fighter pilots, and since we had flown high performance aircraft (the T-38) we were to become their replacements. Please don't throw us into the briar patch!

That took us to Luke AFB and the F-100 school. Due to the shortage, we were given the accelerated course, which we referred to as the crash course. That was six months instead of a year and only 68 hours in the aircraft. Presto! We were instant fighter pilots. But – once again, the fates intervene. There was another shortage; this time it was Class-A Forward Air Controllers. The AF and the Army had an agreement that FACs supplied for the US Army would be Class-A FACs that were fighter pilots and that the Vietnamese Army would get the Class-B FACs that were not. That resulted in our next training in the O-1 Bird Dog at Hurlburt Field, Florida. A slight change from the F-100.

That took me to Vietnam for my first tour. I was assigned to the First Infantry Division at the 2nd Brigade and Division Headquarters, Di An. We lived in tents and flew our Bird Dogs in support of them. I flew 514 combat missions as a FAC plus some more in the Army choppers, but I didn't count those missions, as I wasn't driving. I was awarded the usual array of medals, a bunch of Air Medals, the

Distinguished Flying Cross, and the Silver Star. As you might imagine, I have a few interesting stories from that time, such as the time I dropped a hand grenade into what turned out to be a VC ammo dump - but they are for a different book. When the tour was over, I was told that if I would volunteer for a consecutive overseas tour, they would send me to RAF Lakenheath to fly F-100s again. Once again, I said "Please don't throw me in that briar patch!" and spent the next three years in Jolly Old. When the tour was over, I received the news that the long arm of SAC had pulled the string they had on me and I was going to B-52s at Loring AFB, Maine. Apparently, MPC had my folder specially marked. That precipitated another trip to the O'Club, but for different reasons.

When I got to Loring, I got the usual welcome. Everything was frozen solid. I had not been to Castle yet, so they put me to work at odd jobs, such as investigating a house fire that occurred on base. My first order of business was teaching my wife (who was from the south) to drive on ice. Since the O'Club was closed on Sunday and the parking lot resembled a skating rink, I took her there in our car, a 1971 Pontiac Firebird Trans AM with a 455 HO engine and a 4-speed stick. I first taught her to spin the car and then how to properly correct once the spin had started. It worked and we had no problems except sometimes when the car was so low the snow would pack up under it and lift the wheels off the road.

A few months later they finally sent me to CCTS at Castle. Since I had so much flying time, they could not make me a copilot so I checked out as an Aircraft Commander. That made me one of the few (maybe only) people to be a BUFF AC without ever having been a copilot. The crew that I got when I got back to Loring was made up of whoever was not assigned to another crew at the time. We called ourselves "Catch 38". We had the B-52 G-model aircraft. Since we were the "greenest" crew on base, they did not want to expose us to CEVGs, etc. so when a requirement came up for a crew to go and fly B-52Ds in Nam, they sent us.

So, back to Castle for a D-model checkout and off to U-Tapao and Guam. Since we were flying D-models, most of the missions were from U-Tapao. The ones from Guam were long and boring except one. That night we were scheduled to pick up 90,000 lbs. on a refueling track 20 minutes long. My tanker's autopilot was out, so he was hand flying it. And, oh, by the way, we were refueling in the top of a typhoon. I couldn't figure out why I was having such a hard time hanging on until I looked at the attitude indicator as I was trying to hook back up. We were in a 30 degee bank. He was dodging thunderstorms, but hadn't bothered to tell me. Most of the missions were routine missions in the South, making toothpicks and paranoid monkeys; however, we went north a few times. One time we hit Dong Hoi. Coming in from the sea, I was surprised to see that the town was fully lit up. Nobody had bothered to turn out the lights. Since we were bombing by radar, it really didn't matter, but it was odd. The next time we hit Dong Hoi, on the Fourth of July, we were targeted on the fuel dump. That time there were far fewer lights on. We must have hit the target because as we turned south over Laos, I could look back and see the fuel tanks still exploding. Spectacular fireworks!

One time a SAM came up at us. I went into an (non-standard 80-90 degree bank, fighter style) evasive maneuver and the copilot reported that the missile just "fizzled out" as it passed us. Apparently, the proximity fuse did not function and the self-destruct didn't work either. As we were leaving, the gunner reported an explosion on the ground way behind us. What goes up must come down, I guess.

We finally got a few days off, so our entire crew went to Pattaya Beach, the Thai "Riviera." Since my wife and the Nav's wife were visiting, we rented a bungalow at the Army JUSMAG compound. During the night, a burglar broke in and stole a lot of items including passports, snorkeling gear, and my gunner's underwater camera. We reported the break-in to the front office where we were informed that it had been happening all week. The Officer in Charge of the compound had been staying up all night with his German Shepherd trying to catch the thief, but had been unsuccessful. When we went to bed the next night, I told my wife that he might be back as there were some things he did not get the first time. I decided to set an alarm by tilting a small table against the door so that if the door were opened, the table would fall over and wake us up. My wife scoffed at my "James Bond" scheme, but I did it anyway. At 7:30 the next morning, BANG goes the table. I looked up and there was a well dressed Thai standing there looking surprised. He said, "Sah, you want taxi?" One thing they told us when we checked in was that taxi drivers were not allowed in the compound. I came out of bed, wearing nothing but my red Jockey shorts as he disappeared out the door. I chased him all the way across the compound hollering "Kamoy!" (the Thai word for thief.) He finally stopped and picked up a piece of 2 x 4 and tried to swing it. Since he only weighed about 85 pounds, I ran over him. I grabbed him by the collar and started marching him up to the office. About the time my wife and RN caught up with my pants, I heard him say, "You let me go or I hurt you." I jerked his head around, put my fist about two inches from his nose and said, "Shut up and stand still!" He did and I put my pants on and we resumed our trek to the office. Apparently some of the JUSMAG employees saw me chasing him. They had also seen him get out of a Toyota pick-up earlier, so they called the Thai police and pointed out the pick-up. Shortly after we got to the office, the police came in dragging two more and carrying some of the loot they had stolen the day before. Apparently the compound was under the protection of the King and this was considered an insult to him. Police brutality was not considered an issue in Thailand, and they were treated less than kindly. One lost a tooth. We eventually got back everything except the camera, which they had immediately hocked.

Finally the tour ended and we returned to Loring. Actually it is not true what they said about Loring. They only had nine months of hard winter and three months of poor skiing. If you were lucky, summer came on a weekend. The last months I was there, we had 67 inches of snow. That was between Thanksgiving and Christmas, when winter really set in. The obituaries on the radio were interesting – "Joe Doaks died on Tuesday, services will be at Irving's Mortuary, with internment in the spring." The ground was frozen so hard, they couldn't dig a hole. However, cold storage was not a problem. Neither was chilling a bottle of wine. Just open the back door and stick out your arm into the nearest snow bank. We didn't consider it cold there until you went out in the morning, inhaled, and the snot froze inside your nose. The motto up there was "Don't eat yellow snow!" The social event of the year was "The Moosestompers Ball."

Linebacker II was in full swing and my crew and I were on alert for Christmas. We started our tour on Friday as usual. On Saturday, the Squadron Commander came in and asked "How would you like to be off alert for Christmas?" My reply was "Oh shit!" because we had been following the Linebacker missions and knew of the losses and I knew what that had to mean. Since there were problems at Loring due to rumors, I told the Squadron Commander that if anything happened, I would call him and let him know what happened so he could "short-circuit" any rumors. More about this later.

We departed for Castle for a refresher in the D-model the day after Christmas. I got a new gunner who had heard about the MIG shoot downs and was gung-ho to get one himself. He was cured of that on his first mission. We staged through Guam and while we were waiting to depart for U-Tapao, my RN got sick and my gung-ho gunner had to go through SCAT School since it was so long since he had been to SEA. We arrived in U-Tapao short handed, but flew two "over the shoulder" missions up North with substitute crewmembers. When the third mission came up, we were cleared "solo", and my RN and gunner finally caught up with us. Since we already had subs for them, I asked if they wanted to go. The RN was still feeling puny and said the sub could go for him. The gunner, still being gung-ho, said he wanted to go and get his MIG, so off we went.

Now, Linebacker II had officially ended and bombing had been restricted below Hanoi and Haiphong, but this mission was a 60-ship raid on Vinh. As I remember, we hit with three streams from three different directions. Since all three streams couldn't drop simultaneously, we hit in order with minimal spacing. Guess which position we were in. You got it! "Tail End Charlie" again. Number 60 of 60. By that time, the tactics had been changed to what the fighter pilots had been doing all along. Multiple streams from different directions, etc. That improved things greatly and losses had radically dropped. Another thing that helped reduce losses was that the North Vietnamese had shot up all their SAMs around Hanoi and Haiphong and were flat out. Unfortunately, Vinh still had a full load.

We had support from the Navy and Air Force EW aircraft. They were supplying EW jamming and chaff. However, they did have their limitations. By the time we got there the jammer aircraft had run out of gas and left and the chaff had been blown away by the wind. In addition, we think the SAM gunners didn't try to target the aircraft as they were protected by the jamming and chaff. We think they were targeting the bombs. 108 bombs make one helluva radar return and tell the gunner where to aim. So all they had to do was predict where the bombs were coming from and aim there.

We were doing our SAC evasive maneuver (15° bank and 15° change of direction) on the way in just like good little Crewdogs were supposed to. We were in a left bank portion less than two minutes from bombs away when the copilot reported that he had three SA-2s at three o'clock. I was somewhat preoccupied as I had three SA-2s at nine o'clock. I increased the bank slightly to 85 –90 degrees and pulled on the yoke until the aircraft started to stall and backed off just a hair until the major shuddering stopped. I held the pressure and let the nose sweep down, as I watched the first SAM detonated below us. Then the second SAM detonated below, but much closer. I knew that if the SAM was not moving on the windscreen, it was tracking you. The third was not moving. I applied as much pressure as I dared and said "Move, dammit, move!" At the last second, the SAM started to move down and disappeared under the aircraft. Almost simultaneously we heard a report that sounded like someone had fired a 12 gauge on the lower deck. I reversed bank and turned back about 45° to our inbound track heading and climbed. As we neared our assigned altitude, I reversed bank again and rolled out on heading. I looked at the Time To Go (TTG) meter and it read two seconds. I hollered "DOORS! DOORS!" through the interphone and immediately felt them open and the bombs start to release. Incredibly, we were on track, heading, and altitude (I can't guarantee airspeed). Talk about dumb luck! I checked to make sure the crew was OK and everybody was. We must have hit something as the gunner reported secondaries on the ground as our string hit. We then resumed our evasive maneuvers on the way out. After a minute or two the gunner reported he had three more SAMs at six o'clock. I repeated the previous maneuver even though I knew we had been hit. What choice did I have? The gunner said that the last of the three passed so close, he could clearly see the missiles fins in the glare of the exhaust. As it turned out we had taken hits from three of the nine missiles, one each from the right and left and one from behind. The gunner could see a heavy stream of fuel coming from the left side. After going "feet wet" we did an inventory of the damage we could find. All things considered, it wasn't too bad. In addition to the fuel leak, which we determined was from the left droptank, the autopilot, HF radio, and the gunner's hydraulic system were inop. We later found out that some of the frags had gone into the bomb bay when we still had the bombs. Those were the last nine SA-2s they ever fired at anybody.

From the radio chatter, we knew that one other aircraft, 55-0116, had also been hit and was headed for DaNang. We checked things out and shut down the right drop tank when it had about the same amount of fuel as the left to maintain lateral balance. We determined we had enough to make it back to U-Tapao and we did not want to put two aircraft into DaNang, better known as "Rocket City." Since our evasive maneuvers had separated us from the rest of the stream, we broke off and headed straight home. But the saga continues.

On the way back we kept getting questions from the CP. Some were reasonable, but some were kind of silly. "Why do you have fuel trapped in you drop tank?" "Because it has a big hole in it and the air is leaking out!" As we neared U-Tapao, Charlie Tower called and told us to perform a controllability check and that they were launching a Tanker so we could refuel. I told Charlie to hold off on the Tanker until we did the controllability check, as we might not need him. Charlie replied that it was too late as he was already airborne. Since I was already planning to do it anyway, we dropped the gear, then the flaps and did the controllability check and assured ourselves that all was OK to land. I passed that information on to Charlie and told him we wouldn't need the fuel. Charlie replied that the Command Post said to take on so much fuel in this tank and so much in that tank and orbit until all the other aircraft had landed.

I got to thinking (I do that every once in a while, just to keep people confused.) I didn't know if the gear would come up, and, if it did, I didn't know if it would come down again. The same applied to the flaps. In addition, refueling would put pressurized fuel in pipes that could easily have holes in them. To make matters worse, the amount of fuel they told me to take on would have resulted in the mains being half full. One of the "quirks" of the BUFF was that when the mains were half full and power was applied, the lack of baffling in the tanks allowed the fuel to slosh aft which caused a significant aft shift of the CG. That then resulted in an immediate pitch up. This set of circumstances was most likely to occur if a "go-around" was needed and that was when you really didn't need it. I knew that Ash 1 had crashed at U-Tapao after pitching up on a go-around. I couldn't help wondering if he had refueled and what his fuel state was at the time. Anybody know?

Anyway, in light of all the above, I determined that it would not only be stupid to refuel, but downright dangerous as well. Since my Ma didn't raise no stupid kids, jes' me, I exercised my authority as the commander of an emergency aircraft, told Charlie Tower to "Stand-by", changed channels and landed. Interestingly, although I told U-Tapao tower that I was declaring an emergency with combat damage, there were no emergency vehicles visible on the entire field when I landed.

The 781 write up read "Aircraft hit by surface to air missiles in flight." I figured the maintenance troops needed a chuckle. I never heard my gunner say the word "MiG" again. I talked with the maintenance officer the next day and he said that he had quit counting at 120 holes. The aircraft looked like a Swiss cheese with lot of dents in the bottoms of the wings. Fortunately, no one was in the instructor navigator seat. He would have been a soprano.

I also called the Squadron Commander as I had promised. Later, I found out that the nav's wife had heard a rumor from one of our crews at Guam about our mission. She went to the Squadron Commander and asked him about it. His reply was "I'm not at liberty to tell you." That went over like a lead balloon amongst the crews.

About two days later, I was called on the carpet and a Colonel read me the riot act. He had a "Letter of Admonition" in hand that he wanted me to sign. I reminded him that Air Force regulations gave ultimate authority for an emergency aircraft to the Aircraft Commander, i.e., me, and I refused to sign the letter. He then left the room and returned a few minutes later. He stated that the decision had been made that they could not give me a letter and a medal at the same time. I was instructed to write up the narrative for the medal. Since it was my understanding that previous ACs who had brought back shot up aircraft were awarded the Silver Star, I wrote the narrative for it, and gave it to the personnel people.

A few weeks later, I was told that the General was coming to award medals to me and the AC of the other aircraft that was hit that night. We were to show up at the awards ceremony. About an hour before the ceremony was to start, I received a phone call and was told not to come. The excuse given was that I was on "crew rest". It is my understanding that the Silver Star was awarded to the other AC at the ceremony. When my medal finally came, it was a Distinguished Flying Cross. What do you suppose the chances are that the General would have come to U-Tapao to award two different medals to two people for doing the same thing on the same mission? I have my suspicions and the word "retribution" comes to mind, however I have no proof. Anybody out there know?

One more chapter to this story – after getting back to Loring after the first tour, I heard that they were looking for people to volunteer for SEA tours in F-4s. I applied, but nothing happened. Shortly before we left for the second tour, I decided to call and find out the status of my application. I was told that I had been given the assignment, but turned it down. Since that was totally false, I started asking more questions. Turns out the Squadron Commander had turned it down in my name and never even told me. After some unprintable expletives from me, they asked if I was still a volunteer and I said yes. The F-4s were gone so I took the EB-66 that was left. After that, since I was at U-Tapao, all communication between MPC and myself went direct and not through Loring. The assignment finally came through. Apparently, the Squadron Commander refused to believe it until I called him from McChord on the way home. He was not amused. As we left Loring, we stopped about 100 yards from the gate, opened a bottle of Champagne, poured two glasses, gave Loring a final (digital) salute and drove off.

And so, after 104 combat missions, my saga as a SAC Crewdog ended. I completed EB-66 training only to have them sent to the bone yard before I could get to Korat. I ended up flying O-2s at Osan, Korea, and then becoming an instructor in OV-10s at Hurlburt. After a few more adventures; being involuntarily separated, a 5 1/2 year court fight, reinstatement, working as an aerospace engineer on C-141s and helicopters for the AF, I retired in 1985 at the exalted rank of Captain with 18 years in grade. But, that's another story.

Billy J. Bouquet

My B-52D Gunner
Gary Henley

After completing CCTS as a new "wet-behind-the-ears" EW, in 1975 I found myself assigned to the 20th Bomb Squadron at Carswell AFB, Texas. The aircrews at this "Super Wing" were chock full of expertise, and Crew R-09 was no exception. Capt John Magness was the very experienced and capable AC, and he had plenty of overseas and stateside flying hours to prove it. He was steady with his decision-making, and he ran our crew "by the book." He told us the truth whether we liked it or not, and I could always trust him to make the right decision after weighing all the facts. The copilots (yes, we had two at the time) were Ben Barnard and "Bo" Fulford. I think they were either senior First Lieutenants or Captains at the time. No details ever escaped those two guys, plus they made flying and mission planning fun with their crazy antics. (Note: from a previous volume of this book, you might remember that Ben scored an "HQ" on his initial CCTS check ride as a brand new AC).

The expert talent up front in the cockpit was matched downstairs with the Radar and the Nav. Capt Dick DeRoos was the Radar, and Capt "J.T." Turner was the Nav. Both Dick and J.T. were really good together, and we never worried about bad bombs with that dynamic duo. I met Dick six years later at SAC HQ where he was working in NR (Science and Research). Dick was a real genius, and SAC benefitted from his great ideas at the MAJCOM level. Rich Vande Vorde (the Radar Nav who took over when Dick PCS'd from Carswell) told me recently that most people don't realize this, but from the Initial Point (IP) to the target, J.T. was as solid as a rock.

If the skills of the aforementioned crewmembers weren't daunting enough, our gunner was Airman First Class Billy J. Bouquet. In my entire 30-year career in the Air Force, I never met any airman who was sharper than A1C Bouquet. He was astoundingly smart, knew his job inside and out, and helped improve the morale on the crew with his professionalism and with his talents in music. (See lyrics to a few of his songs later in this volume.)

We always maintained a professional officer/enlisted relationship; however, there was often the bantering that goes on between aircrew members and, after all, he and I comprised the defensive team. I remember a particularly long mission planning day when we had changed our refueling route twice and were re-accomplishing our paperwork for the flight. Keep in mind that this kind of mission planning was long before personal computers, and I couldn't find my pencil. Yes, we did most of it by hand in pencil or ink. I recognized the color of my pencil in the "Guns" hand and asked him to give it back to me so I could finish my work. He remarked, "What makes you think this is your pencil, Lt. Henley? Is your name on it, sir?" Thinking that he had really put me on the spot, he grinned at me, while I replied confidently, "As a matter of fact, yes, it is! Take a look!" Sure enough, I had bought a bunch of pencils that month from a Walter Drake catalog, and had "Lt. Gary Henley" embossed in silver color ink on each pencil. He almost fell right down on the floor in shock! The rest of the crew got a huge laugh out of it because that was the only time anyone had ever got the best of our Guns. You can be sure nothing like that happened, again - Guns was that sharp!

Among the huge numbers of Southeast Asia veterans in the squadron and on the base, there was an amazing variety of souvenirs that had been brought back from overseas. Among these were "Papasan" chairs, wicker furniture, hand-made wooden model airplanes, fine china, and pachinko machines, to name a few. Pachinko machines were fascinating to me because I had never seen anything like them before or since in my life. They were a strange cross between a pinball machine and a slot machine, except that there was no electricity involved - just gravity. I think they got the name because of the sounds made by the ball bouncing along inside made noises like "pa-chink-o." The way I remember it, a pachinko ball that looked like a large metal ball bearing would be inserted and with a thumb action on a lever be shot to the top, and then it would bounce along metal posts and levers as it made its way to the bottom of the case. Depending upon where the ball landed, it could release other balls that had previously been dropped into the machine, resulting in several of these balls cascading down the passages and levels of the machine. The object was to get more pachinko balls out of the machine than you put into it, and then you redeemed the pachinko balls for cash or prizes. I wonder how many guys still have pachinko balls in their pile of souvenirs. The only reason they stand out in my mind was that the environment for a pachinko ball was similar to that of a B-52D tail gunner.

I always considered the B-52D gunners as a special caliber of aircrew members. They went through more rough flights than even fighter jocks. Fighter jocks had nothing on our gunners. I'm absolutely certain that our tail gunners had cast-iron stomachs. Gunners had to have cast-iron stomachs because their work environment was terrible! Consider their pressurized compartment at the end of the 180+ foot fuselage. In turbulence, when the front compartment moved two feet up, the other end moved a factor of about four the opposite direction (i.e. eight feet down). Positive G's ended up front were negative G's in the tail, and vice-versa. Left and right movements also were exaggerated. On quite a few flights, Billy Bouquet's flight helmet (the old white plastic kind of the 70's) was completely busted and useless. He had to have another new helmet issued. Note: this occurrence was not uncommon among the gunners, and I remember that ours was reissued two new helmets just in the time I was on the crew. Those gunners had to literally hold onto the handles around them for their dear lives or have their head or helmet bounced around like a pachinko ball. Air turbulence is annoying enough at high altitude when you least expect it, either in severe weather or in mountain effect or jet stream effects. In our case, however, we routinely practiced penetrating enemy airspace at low altitude to avoid detection by air defense systems.

When you mix low altitude tactics with hot air and/or thunderstorms, look out! It can get pretty vicious back in the tail section for the gunners. On several suspiciously fortuitous preflights, Billy had the uncanny occurrence of a "bad inertial reel" in the aft compartment that ended up with Fire Control having to come and install the stow pin for the aft section because they were unable to fix his inertial reel prior to takeoff. He would then either fly with the rest of the crew up front or just go home for the day/night while the rest of the crew flew. Perhaps some day he will explain his "technique" since the "statute of limitations" has long expired. (See the song Billy Bouquet wrote in this volume called, "I'm Not Flyin' Tonight.")

Another difficult situation for gunners involved a dangerous condition when the pressurization would fail in the aft section while in flight. This would result in the crew descending the aircraft to below 10,000 feet for the gunner to crawl along the 14-inch crawlway from the end of the airplane to the front compartment of the airplane to join the rest of the crew. Note that this had to be accomplished with his helmet and parachute on! Billy had to do that on more than one occasion. I thank God that I never had to do it. To put this situation into perspective, imagine a B-52 sitting on a football field with its tail bumped up against the goal posts, its wings hanging well over the sidelines, and its nose at about the 50-yard line. This would mean that the gunner would be crawling along this crawlway from the goal posts all the way to the 50-yard line.

What's my point? I firmly believe that the gunners had it unusually tough in our airplane, and I would like to thank them for what they endured to keep us safe from enemy fighters during wartime, as well as friendly air traffic in peacetime. The B-52D had a fantastic fire control system (FCS) that I wished some gunners would tell us about in this series. One thing that I remember is that once an object was tracked inside the FCS's "cone of death," there was 100% chance of shooting it down. During the Vietnam War, two of these guys were MiG killers. Aircraft 55-0083 at the Air Force Academy is a testimony to this fact about those important airmen. Billy and I had quite a fun time as we flew Fighter Interceptor Exercises (FIEs) against the Air Defense Fighters from Houston out over the Gulf of Mexico. It was only on those flights that I was allowed to dispense flares or chaff and to actually practice jamming against a live "threat." Those F-106s and F-4s were quite a challenge to shake-off when you consider the maneuverability of the B-52 and the huge radar cross section we were trying to hide behind jamming and chaff. Thankfully, we had a pretty good track record of breaking their radar track lock-on with our chaff, jamming, and maneuvers. That gave us some confidence, considering the state of the art in our adversaries' systems.

I finally got to read about some of the gunners' perspectives in the first volumes of *"We Were Crewdogs,"* and I hope we hear more from them in the future. It seems that the only stories you hear are the buffoonery antics for which Crewdogs are well known. For instance, in 1975 there was the new gunner on alert that really didn't understand the concept of "Seven-Day Alert," and one day was caught driving the crew Alert truck through the main gate at Carswell AFB. You see, he thought that he could take the truck home each night, then bring it back next morning prior to the morning alert briefing. Nobody on the crew realized that each night their gunner was missing in the alert shack. A few heads rolled regarding that incident because he wasn't caught until halfway through the alert cycle. Nobody had ever made it clear that he was expected to spend each night at the alert facility! Remember this story the next time you are training someone to do a task—do not assume anything. Spell out everything as if you were explaining the task to your mother. There probably are a thousand untold stories just like this one out there that most people are embarrassed to tell. We'd love to see them all in print for a good laugh and to prevent future occurrences.

You may ask, "Whatever happened to Billy Bouquet?" Well, about six months after I left the crew to go to Stan/Eval, my defensive team "wingman," Guns, finished his four-year enlistment in January of 1978, whereupon he departed the Air Force as Sgt. Billy Bouquet. He returned to his home near Houston where he resumed his old job in the grocery business and where he is now a very successful businessman I'm not surprised in the least. Someday, I hope we can have a crew reunion somewhere. Until then, I'll just have to remember my gunner and those other B-52D tail gunners who were among the "rough riders" who withstood some challenging low-level training route turbulence in the late Seventies.

They were a rare breed of airmen!

My Encounter with Immortality
Robert O. Harder

It was an exciting time in May of 1968. I had just arrived at McCoy AFB in Orlando, Florida, after three months of B-52 "finishing school" at the 4017 Combat Crew Training Squadron (CCTS), Castle AFB in Merced, California. There I was awarded my new Air Force Specialty Code (AFSC), 1525D, B-52 Navigator. During earlier Navigator-Bombardier Training (NBT) at Mather (a.k.a. "Mother") AFB, California, I had been schooled in the Q-48 Bombing and Navigation Systems (B-52Cs and Ds), and McCoy was one of only eight bases still equipped with those older "Big Belly" BUFFs. The timing of my arrival at that particular station was not accidental; there would be just enough weeks for my new crew and me to be checked out and certified Combat Ready before the 306th Bomb Wing rotated back on Operation Arc Light.

That was fine. As a no-strings bachelor First Lieutenant with a yen for a little adventure, I was actually excited about the prospect of going to Southeast Asia. Despite the long and fatiguing three-day cross-country auto trip, I was still pretty hyped that Sunday evening when I drove up to McCoy's front gate off the Bee Line highway. I showed my PCS orders to the guard, got my first salute as an "operational" Air Force officer, and headed for the BOQ. After checking in, unpacking, and getting my head straight regarding reporting in to the 367th Bomb Squadron the next morning, I was suddenly overcome by hunger and thirst. Mostly thirst.

It was off to the O'Club then, a short stroll from the BOQ, arriving about 7 p.m. After a quick dinner in a very quiet and sparsely seated dining room, it was definitely time to find the water hole. My destination was but a few steps down the hall.

I opened the door and tentatively peeked into what, at first glance, appeared to be a completely empty, darkened saloon. My heart sank. Could it be they were closed?! A moment later, someone appeared from a room behind the bar, turned on a few more lights, and motioned me in with an arm wave. Much relieved, I hurried over to the bar and parked myself on the closest end stool.

"Sundays are always slow," the bartender explained in response to my unasked question. "We're usually locked up before nine." He had on civilian clothes, but bore a strong military look and bearing; rightly or wrongly I assumed him to be a moonlighting noncommissioned officer.

"Glad you're still here," I said, and meaning it fervently. I had been thirsty earlier, but the unexpectedly sudden prospect of being deprived entirely had brought on a case of dry-mouth worthy of one of those old-west bullwhackers crossing California's Death Valley.

The sergeant smiled indulgently. "What'll you have, Lieutenant?"

"Scotch and water," I said, naming the rookie USAF flying officer's drink of choice.

He ostentatiously poured a stiff one - to my immense pleasure - and set it down in front of me. I brought the glass up to my lips and was about to drink deeply of the golden nectar when from the other end of the bar a glass slammed down hard on the counter.

"Hit me one more time!" came a booming, authoritative voice.

The sudden outburst had been so startling that I missed my mouth, dribbling scotch on the bar counter top and down the front of my Class-A Blues. My eyes still weren't adjusted to the darkened clubroom and I hadn't noticed another person at the far end of the bar, partially hidden behind a cluster of tap beer dispensers.

The other guy peeked around the beer taps and stared curiously at me, as if examining some newly discovered and exotic animal species, while the bartender hurriedly poured his refill. Grasping the just delivered fresh drink, he stood up from his bar stool and, a bit unsteadily, began to work his way over toward my end of the bar.

"Hi ya, Lieutenant," the fellow said not unkindly, as he plunked himself down on the stool next to me and half-drained his glass with one swig. There was just a hint of a slur in his voice.

I gulped hard. The shirt collars on his khaki 1505s were adorned with eagles.

The colonel glanced at my shiny silver bars, slick navigator wings, and deer-caught-in-the-headlights expression, drawing the logical conclusion.

"First assignment out of Castle?" he asked.

"Yes, sir," I said in my by then well-honed, highly deferential Air Training Command response voice.

He nodded, sipping slowly at his drink, his eyes fixed on a spot above a double row of bottles lined up on a shelf behind the bar. I unconsciously did the same thing, silently cursing my bad luck. Instead of being able to let my hair down and suck mindlessly on my scotch, I would now have to keep my wits carefully about me. Ye Gods, I had never even spoken mano y mano with a full colonel before!

"Tanker or bomber nav?" the colonel asked conversationally. McCoy also had a KC-135 Stratotanker squadron.

"Bomber, sir," I said, trying to clear my throat of a large lump. All I needed now was to put my foot in it by saying something stupid to a high-ranking wing officer. "I'm reporting in tomorrow morning," I plowed on. "Guess either I'll replace another nav on a current crew or they will form a new crew."

My voice trailed off uncertainly; I really didn't know what I was talking about - just repeating some half-formed scuttlebutt I'd picked up along the way. Increasingly, all I could think about was how to get away from this guy and return all my attention to the scotch.

"You'll be on a newly formed crew," the colonel said knowledgeably, after one more glance at my nametag finally seemed to have registered with him. "We need to get several additional Combat Crews ready over the summer, before the wing rotates back to SEA."

I voiced another respectful "Yessir," while surreptitiously checking out the cat's Master Navigator/Observer wings and ribbon rows. The old boy had been around.

"They're still cross-training you fellows out at Mather, right? Gotta be a nav before bomb school?" He knew the answer of course; the question was a conversation starter.

I nodded and was preparing to blather on about how wonderfully I'd been educated in Air Training Command and how eager I was to become an elite SAC Combat Crewman when the colonel mercifully interrupted me. He seemed to have something specific on his mind.

"So, tell me, what do you think so far of the Air Force?"

I reflexively spit out an answer, without giving any real thought about the oddness of such a question coming from a senior officer. "I have a Regular commission, sir," I said formally, "I plan on making the service a career." I was proud of how firmly I had stated that last.

The colonel looked at me hard for a moment, and then studied his drink glass. "Well," he said quietly, "I hope it works out for you." He paused for a long sip, as if uncertain about whether he should say anything more. The too-much-to-drink unsteadiness I'd observed earlier had disappeared. At length, he finally gave in to his thoughts. "Look, let me give you a word to the wise—you would be smart to get out of the navigator and aircrew business as soon as you can. Transfer into a service or support organization and make your mark there. That is, if you entertain any thoughts about rising to any serious rank in this man's Air Force."

"Sir?" I was puzzled, if not a little shocked, by the remark. What could be better for a prosperous and lengthy career than being a rated flying officer? Why would I want to get into, say, personnel or supply or maintenance or whatever, when the cockpit was where all the action was? And why was a senior officer planting such ideas into the head of a lowly First Lieutenant he'd only just met?

"Well, maybe I've said too much," the colonel said, seemingly reading my mind. "But the fact is I'm retiring pretty soon and I'm done with all the standard U-Saff horse crap about Mother, God, and Country. These days, I pretty much call 'em the way I see 'em." He turned on his stool and looked me in the eye. "This is a pilot's Air Force, kid. It's only the pilots that get the good operational jobs, all the commands. And if you think you can get around that in mid-career by cutting laterally into one of the non-rated, support fields, forget it. They only promote their own guys, those that have worked their way up through each organization from the start. Fact is, in the end, the career non-pilot Crewdogs - the navigators, bombardiers, and EWOs - get diddley-squat." He touched my arm, looking almost sorrowful. "You won't like this but I can tell you right now what is going to happen - you are almost certainly going to be in SAC your entire career. After three or four years as a nav, you'll upgrade to radar nav, a job you will keep for at least the next ten years, and probably longer. If you're clever or lucky, you might some day wangle a way to unstrap the beast from your ass and finish with a job in the command post, wing bomb/nav shop, or maybe somewhere in staff war plans. For sure," he tapped the eagle on his left collar with a forefinger, "you can forget about wearing one of these."

With that, the colonel bottoms-upped his glass and thunked it down in a gesture of finality. "Like I said, Lieutenant, if I were you I'd just chuck the whole thing - get the hell out at the first opportunity." He stood to leave.

I wanted desperately to ask him a host of questions: 'I don't understand, sir. What about you? You're a nav, just like me, and judging from the badges and salad on your chest you've had a great career and made full colonel. It all worked for you, didn't it? Why wouldn't it work for me too? How come I don't have the same chances as you?' I wanted to ask him those things, and much more, but I never got the opportunity. He was already striding out the door, tossing a farewell finger wave at the bartender.

I turned back to my scotch, exhaling hard in relief that at the very least the weird encounter was finally over. I took the balance of the liquor in one large gulp and signaled for another round, saying to the bartender, "Boy that was really something!"

The sergeant came over with my refill. He had heard every word of the exchange. "Lieutenant, you got any idea who that was?"

"No I do not," I said, working myself into a bit of a snit. The old Colonel had gotten to me more than I realized. "And what's more, I don't think I want to know!"

The sergeant thought differently. He leaned into me, both elbows on the bar, his face strangely serious. "I'm guessing you'll probably change your mind after I tell you. He's Colonel Tom Ferebee, bombardier of the Enola Gay, and the man that on August 6, 1945 dropped the atomic bomb on Hiroshima."

Satisfied with the shocked look on my face, he dragged his bar rag through the scotch puddle in front of me and promptly disappeared into the back room without another word.

I stared at my fresh drink for a full minute before finally pushing it away. I stood up, tucked my blue service cap under my left arm, and departed for the BOQ. Sleep came hard that night.

Colonel Thomas J. Ferebee retired to Windermere, Florida, in 1970 after 28 years of service in the U.S. Army Air Force/United States Air Force. Ferebee was already a veteran combat bombardier with 60 missions in Europe when in 1944 then Captain Paul Tibbets selected him to become part of his specially formed 509th Composite Group, scheduled to be equipped with the new B-29 Superfortresses. By the time of the big launch from Tinian on 6 Aug 1945, then Col. Tibbets considered then Major Ferebee "the best bombardier who ever looked through the eyepiece of a Norden bombsight." Ferebee later served in B-47s before ending his career with the B-52Ds of the 306th Bomb Wing at McCoy AFB, where the above true incident occurred. Colonel Ferebee, age 81, passed away at his Windermere home on March 16, 2000.

The Enola Gay has recovered from attempts in the 1990s to smear her reputation and is proudly receiving callers in the Smithsonian National Air & Space Museum's new Udvar-Hazy facility at Dulles Airport, Washington DC. Visitors can peer up through the bottom of Ferebee's "green house" and see Norden M9B Bombsight Number 4120, the very device he used to drop "Little Boy."

Bob Harder's new book is titled *"Flying from the Black Hole: The B-52 Navigator-Bombardiers of Vietnam,"* published in hardcover by the Naval Institute Press, Annapolis, MD. Available May, 2009. List Price $34.95. For more details and online discount purchasing information, check out his website: www.robertoharder.com The book can also be directly ordered from any bookstore in the U.S.

The Flight of Tuff-13
Stephen Henley

I'm a recent Electronic Warfare Officer (EWO) graduate of the Formal Training Unit (FTU), the 11th Bomb Squadron. It is somewhat the equivalent of Combat Crew Training School (CCTS) back in good ol' SAC days. For those interested in the training process, the entire B-52 operational community is currently trained in one unit, though that training is actively being augmented with the 93d Bomb Squadron, the only Air Force Reserve unit that flies the mighty BUFF.

In each class, the navigators are the first to arrive, and they quickly delve into academics. Shortly after the navs come the pilots, who also start their academics. After about another month, the EWOs join the class. The EWOs spend a lot of time in the T-4 simulators initially, and they actually complete the first half of their checkride without setting foot inside the jet. After academics, all the students spend some time in the Weapon System Trainers (WST) learning to work together with their respective "hard crew." Each person is assigned to one crew, and that crew stays together throughout the entire training. In the WST, the EWOs get stuck in a box that doesn't move which has an ancient and faint microphone connection to the rest of the crew. The navs get in a 1D motion simulator with significantly better communications, and the pilots get the "Gucci" - a full-motion and video simulation of 1960s technology approaching the year 2010.

After doing some crew coordination training as well as egress training on what's left of a B-52G's nose section, the students hit the flightline and fly their first sorties. During these flights, we learn aerial refueling skills, bomb run procedures, defensive maneuvers, various radio communications, and apply the knowledge learned from academics to real aviation. Of course, this includes the inevitable—we beat up the pattern like our predecessors did…many, many, many times. Despite the EWO queasiness that comes with flying backwards in a dark, hot environment with the newest BUFF drivers, sometimes it's good not to have a window (see following picture)

After about 100 flight hours, we get a checkride and are shortly thereafter certified as combat crewmembers and ready to join the Combat Air Force. Here's my story:

When I arrived at the 11th Bomb Squadron, due to a SNAFU with my orders and out-processing from Randolph AFB, I arrived approximately three days after my class start date. I was given the option of starting my training and playing catch-up or waiting for an additional three months. I had no desire to sit around doing nothing, so I elected to play catch-up. I started my training with a simulator tour to learn more about my station. From there I spent the next 2½ months learning the EW systems, learning how to employ them, practicing in the simulator, and focusing in general on how to be a good crewmember. After this academic period we were sent to the "flightline" side of the house. We were told that sometime during our flight curriculum one crew out of our entire class might get the chance to drop an inert round at the end of the training course. Imagine our surprise as we neared the finish of our training when we found out that our crew actually would get to do a live drop of 27 Mk-82s (500-lb bombs) prior to our checkride!

We started the mission planning for it with a basic route study followed by target study. After that, we went over the plan for what we'd do in the case of something going wrong. Among other topics discussed was what to do if all the weapons didn't drop out of the bomb bay. If we dropped the ordnance and the spotters on the ground didn't count all the bombs impacting the ground, we'd have to do a bomb bay check, a process by which we had to descend below 10,000 feet and a weapons-qualified Crewdog would have to weasel through the narrow 14½-inch catwalk back into the bomb bay and make sure all of weapons came off the jet. During our three passes of nine bombs each, there was plenty of opportunity for problems, but nothing we didn't think we could handle. We briefed the rest of our flight and went home for crew rest.

The next day, we arrived, got our gear, got a step brief, and preflighted our 49 year-old, H-model BUFF. We even took off a few minutes early and began the long trek from Barksdale to the Utah Test and Training Range (UTTR...for those not familiar with it, it is pronounced You-Terr). Enroute, the headwinds weren't as bad as we thought, and we ended up being quite early to the range. In order to kill some time before entry, we decided to take a tour of the Grand Canyon...

Great view...bad idea...

The spectacular view turned out to be a curse as the air traffic controllers (ATC) delayed our turn north towards the UTTR for about 20 minutes because of traffic coming from/through Las Vegas. This frustrated us as we had a limited amount of time in the area. Once cleared by ATC, we headed straight for the range and got there about three minutes late, but found some F-16s still in the airspace during our scheduled time. I don't envy the controllers at the range and their balancing act between two flights in their airspace, but our time was limited as it was. Our IP was irate that another delay was minimizing our time to drop live weapons (a rare training experience).

We opted to go ahead and fly towards our area of the range and avoid the F-16s' area. As we rolled into our first pass on a dry run, the F-16s left the area, and we finally got the clearance to do our first drop. Our drop areas for chaff are pretty limited, so when I got the chance to punch out a few bundles, I dropped them out into our designated chaff drop area during the post target turn. Since it was so rare to have the chance to drop chaff, my instructor asked if he could punch some out on the next pass. I said, "Go ahead."

From the dry run, we went around and went in on our first live bomb drop. Everything went perfectly—we were on time, on centerline, on airspeed, on altitude, and every indication inside the jet showed nine weapons dropped towards the target as we made our break-away.

My instructor hit the button to punch out some more chaff as we awaited results from the ground controllers. Meanwhile, the offense closed the bomb bay doors, the aircraft shuddered, and I heard a grinding sound. (Normally, you barely feel the doors open and close, if at all. To feel or hear anything other than a small thump is quite an anomaly.) The offense announced that they weren't able to latch the bomb bay doors and tried to manually cycle the doors from closed to open to closed. Unfortunately, that didn't work, and the same grinding sound and shudder re-occurred.

Meanwhile, I looked over at my chaff counter as saw the numbers on my chaff counter continuing to count down well past what I programmed them for no apparent reason. I immediately halted the system and pulled the circuit breaker, figuring I'd deal with it after we diagnosed the bomb bay issue. Later, the RN said we punched out so much chaff that the radar image looked a lot like someone drew the letter "J" over our route on his radar scope!

In the midst of diagnosing this bomb bay issue and the chaff problems, the ground controllers called in and told us, "TUFF-13, TOSS. We've got seven good spots."

We on-board indications showed that we released nine weapons, so the IP queried, "Can you check again?"

"We've already checked three times just to be sure, but we'll review the video again."

"Thanks."

After a short pause, they responded, "TUFF 13, TOSS still only counts seven impacts."

With the guys on the ground witnessing only seven impacts, in the back of my mind I'm imagining a 500-pound Mark-82 dangling from the internal racks and being squeezed by some VERY powerful B-52H hydraulic doors.

So, that means we're going to have to check the bomb bay. Well, that's just great…now we have two problems that compound each other. So, we descend to below 10,000 feet (Did I mention that the altitude of some of the terrain in the bomb drop area was well above 10,000 feet?) and the IN crawled back to the bomb bay on that 14½-inch catwalk. Due to the fact that we could not guarantee that the doors were secure, he opted to peek just his head into the bay without going into the bay itself. While doing that delicate maneuver, he had to wear his parachute and helmet. With the landing gear retracted, that left almost no room to maneuver.

So the IN checked and came back to tell the rest of the crew, "Well, I have some good news…and some bad news. The good news is that we did drop nine weapons. The bad news is that I just had a great view of the mountains in Utah through the bomb bay."

So, here was the complexity of the problem. It really isn't preferred to have a B-52 return to base with unexpended weapons; however, we were having problems with the bomb bay doors.

We first considered jettisoning the weapons, going so far as to talk about the point at which we would jettison them. During a rather lengthy ensuing discussion, the IN and nav team realized that merely jettisoning the bombs directly over the target actually would result in the impact of the remaining bombs several miles beyond the intended impact point. We then considered that no matter where we jettisoned the weapons, Explosive Ordnance Disposal (EOD) personnel would be required to disarm 18 Mark-82s, potentially risking the lives of more people. This was a "real world" situation, so we heavily pondered that option. Meanwhile, someone pointed out that we don't have a problem with the weapons, but only an issue with closing, but not opening, the doors. Since nothing appeared wrong with the bomb release system or opening the doors, we opted to drop our remaining weapons on an alternate target.

So, we set up, rolled in and dropped our remaining 18 bombs—everything seemed to run perfectly. To our surprise, the bomb bay door also closed properly.

The controllers then called and told us, "We've got 14 good spots."

The frustration and tension was evident from the pilot when he asked through clenched teeth you could hear on the radio, "Can you check again?"

"Uh, yeah…no, make that 16 good spots."

Oh, like that makes a difference, I thought to myself.

So we were obligated again to descend to 10,000 feet and perform another bomb bay check. Fortunately, we confirmed that all the weapons came off the jet, again. Next, we let the range know that we were WINCHESTER (all ordnance expended) and then left the area.

After four hours of flying back to the base (to include an hour and a half of night pattern work), and we finally set foot on "terra firma" for a few hours of mission debriefing. By the way, later in training, we also got to drop M-117s, as well as inert JDAMs before our checkride; however, none of these drops compared to that first experience.

From the previous volumes of *"We Were Crewdogs,"* how those SAC crews flew those 10-12 hour missions after a maintenance hold on the ramp for many long hours prior to take-off is a significant testament to their stamina and training, because I was wiped out after this "mere" 8.5-hour sortie.

My Mother Never Understood
Tommy Towery

When I graduated from college with a journalism degree, my mother knew that I might one day write stories for a living. She understood that. When I elected to avoid that profession and instead go into the Air Force through the ROTC program and opted to go to pilot training, she understood that I might someday fly airplanes for a living. My failure to reach pilot status directed me toward navigator training which could have led me to a living directing pilots where to take the airplane – that was also understandable, even by my mother. Then I muddied the water. When I continued my advanced Air Force flight training by entering into the world of electronic warfare and to my new chosen profession as an Electronic Warfare Officer I entered into a black hole where nothing I learned or did could ever be understood by my mother or by most of the other people who knew me. It was easier just to tell them I was a Crewdog!

A popular web based dictionary defines Electronic Warfare (EW) as "*the use of the electromagnetic spectrum to effectively deny the use of this medium by an adversary, while optimizing its use by friendly forces. Electronic warfare has three main components: electronic support, electronic attack, and electronic protection."*

Boy, now that makes it simple! "Get with it Mom!"

The mascot, if you will, of the EW training and the profession is a black crow, usually holding lighting bolts. According to the Association of Old Crows, *"During World War II Allied ECM officers, tasked to disrupt enemy communications and radars, were given the code name of "Raven" to provide a degree of security to their existence. After WWII, a group of Raven operators were directed to establish a SAC flying course in ECM operations at McGuire AFB, New Jersey. From all accounts from those present at the time, the students changed the name to "Crows" and those engaged in the profession became known as Old Crows. "*

The current EW training squadron offers the following information about what is required to become an Electronic Warfare Officer:

An Electronic Warfare Officer is a rated navigator with advanced training in enemy threat systems and electronic warfare principles. An EWO identifies and counters weapons systems such as: Surface-to-air missiles (SAMs), Anti-aircraft artillery (AAA), and Enemy fighters (AIs).

I hate to admit it, but that sounds simple enough to understand, at least to me it does. Of course my judgment is clouded by years of dealing with those factors on a daily basis as well as finding myself surrounded by others who had chosen the same profession as I did. I guess I could have told my mother that I identify threat weapon systems and counter (reduce the effectiveness of) them. How I do that does not really matter to most of the world, not even the fellow Crewdogs who occupied the B-52 in which I was flying, let alone my mother. And who wants to tell his mother that there is some enemy trying to shoot his airplane down? Just do it! Simple!

Every generation of EW has suffered his own agony of a training program I am sure. By the time I began my departure to the dark side in 1971, the personal computer was still unknown and even the integrated chip was in its infancy. It was so rare that we almost ignored it and

89

instead spent many months learning about vacuum tubes, resistors, capacitors, and were marveled by the tiny transistor. My class had to learn to read the color codes on resistors using nice little saying like "Bad boys rape our young girls but Violet gives willingly." which was a useful mnemonic for remembering the first ten color codes – black, brown, red, orange, yellow, green, blue, violet, gray, white. We studied the basics of vacuum tubes, since they were still around, and the inner workings of transistors. We not only studied them – we were tested on them.

Our EW instructors tried to make things interesting, even though most of us held liberal arts degrees and there were few engineers or rocket scientists in our classes. One instructor managed to mesmerize us by standing under the black light at the front of the classroom and juggled the fluorescent chalk that was used to draw circuits on the blackboard. We were easily amused in a new and complex electronic world. After all, it was the Sixties and black lights were far out.

Many of us were exposed to our first classes in working in a secure building, where we had to show badges to a guard to gain entry and learned not to leave the badge at home on the dresser or kitchen table. That training would come in handy later in life when we learned about "alert." We studied classified things from classified text books which could not leave the building – a situation that required us to make special trips on some evenings just to study for tests. We listened for hours to the audio tapes of harmonic beeps and tweeps of various radar systems to learn the unique characteristic sound each made and to identify the type of threat based on those sounds. We found that Surface to Air Missile (SAM) systems sounded differently than long range defense radars. Fighter radars had various sounds for search, acquisition, and lock-on. While the color codes of resistors would mean little to us in our later lives, we knew immediately that the radar sounds could mean the difference in life or death. We studied those a lot.

We learned about chaff - the shiny strands of aluminum foil that could be ejected from an aircraft to create a false target for radars and found that it was cut to specific lengths, based upon the radar system it was designed to counter. We learned about long strands of rope chaff and how we never wanted to accidentally drop it over the United State, lest we blackout a whole city by shorting out the high-tension wires.

We learned about flares, and how they had been designed to offer a higher heat-related target than our own jet's exhaust to an enemy weapon system and how they could alter the path of an enemy fighter's heat-seeking missile to save our aircraft.

We learned about warning receivers and jammers and antenna makeup. We learned lots about antennas. There was the stub, the blade, and the phased array, parabolic "dish", the scimitar, yagi-array, and others. We learned about frequency bands used by these and the weird names they were given. Echoes of Ku band and Echo band and Sierra band haunted us in our sleep.

Besides the classroom instruction, my class flew missions in the ET-29D, a modified T-29 Flying Classroom. It was still the twin-engined Convair, but it had some receivers and jammers installed instead of the Flying Classroom's normal navigation training gear. We even got to jam some real radar (with prior ATC approval) on some of the flights.

We took classroom tests and inflight examinations and worked together hard to get through the process of becoming Electronic Warfare Officers. There was healthy competition among the classmates, because the choice of assignments were offered to the graduates based upon class standings, and the highest achiever in the class got the first pick of the assignments handed down to the class. The next highest went next and on and on down the line. My class worked especially hard because there were rumors that back-seat F-4 Phantom slots were going to be available and maybe some gunships and Wild Weasels could be had by the best of the best.

We also knew there would be lots of B-52 assignments, and, as much as I hate to admit it, none of us wanted to go to B-52s when the chance for a fighter plane or a combat aircraft was possible. The idea of sitting on alert 120 days a year and flying nothing but training missions was unthinkable to a bunch of new Lieutenants who wanted to go out and kill Commies. This was 1970 and the war was still going on. I now wonder in reflection why no one mentioned the phrase "Arc Light" to us.

When the time came and the assignments were posted, the eight members of my EWO class waited with baited breath. We clamored to see the posted list only to find out that the fast-movers and Commie killers promised to us were no where to be found. Eight students, eight Lieutenants, eight graduates, eight eight-engined B-52 assignments! The only reward for being the best of the best was getting the first pick at which base's B-52s we'd be flying in. My good friend Bob Pettit was a bachelor from Connecticut and was one space higher in standings than me. He knew my ex-wife's parents lived near Ft. Worth, so he took Dyess and left Carswell for me – the next in line. He didn't know how out-of-place a Yankee from Connecticut would be in Abilene, Texas – especially one who drove the only lemon VW ever sold – a Karmann Ghia. Did I mention he was Catholic?

Following a trip through the B-52 Combat Crew Training Squadron at Castle AFB, California, flying B-52F aircraft, I was on my way to Carswell. I flew three upgrade missions there before I was grounded for kidney stones – two missions in "C" models and one in a "D" model. It took two years for me to get back on status and become a real Crewdog.

During the time I was grounded, the skills learned about the resistors, capacitors, and transistors were put to the only practical use I ever had for them. This was not the result of any needs of the Air Force. I never touched an internal component of any black box as a crewmember – they were always "remove and replace" items. If one of them didn't work I just wrote it up after the flight and they (the maintenance folks) jerked it out of the console and put in a working model so we could get the ship off on time. They should have spent all that time at EW school teaching us penmanship. No, I used the electronics knowledge hammered into my brain at EW school as many of my fellow EW officers did – building a TV set from a kit.

Whether it was great word-of-mouth, or an aggressively savvy marketing campaign, the DeVry Institute of Technology had many Electronic Warfare Officers and some other crew positions sign up for their Home Electronic Technology correspondence school to learn how to repair televisions. The $1,250 school tuition was 90% paid for by the VA, and after completing 105 lessons, the graduate not only ended up with a framed certificate, he also had a 25-inch color Zenith television which he had built from scratch - assembling circuit boards of resistors, capacitors, transistors, and other pretty little electronic thingies. The picture in my completed TV set was upside down the first time I turned it on since I had somehow installed the yoke upside down. During the course I also built a voltage meter and an oscilloscope. Many alert tours were filled with nights of working on the mail-in (long before online courses) lessons in the big orange three-ring binders, and I personally ended up taking many of my lessons with me to Guam during "Bullet Shot" since you had to complete the whole course in a year's time. The kicker was that you didn't get the TV kit until the year was almost up, so if you dropped out, all you had to show for your money was a few paper lesson books and maybe an oscilloscope. I figured if I got out and became a TV repairman, my mother would be able to understand that.

While I was fighting the medical boards to get to return to flight status, I worked a while in Combat Intelligence. It was easy for my mother to tell folks her son was an Intelligence Officer. Most people had heard of that before, but did not necessarily know what one did. It sounded impressive though. When I finally won my battle with the medical board, the powers at be decided that I needed to go back through CCTS, since I had never really qualified in a D-model after my schooling at Castle, and all the C-models had been retired. The 4018CCTS was just forming at Carswell, and I was sent back to "basic" training there. Once again I learned to monitor warning receivers, mark signals with an orange grease pencil on the large orange scope in front of my position and how to center jamming packets on top of radar signals displayed upon it. I learned how to monitor two UHF radios, two VHF radios, one HF radio, and one warning receiver and still communicate with five other crew members at the same time. I also learned my part in the nuclear two-man plan by being able to pull and stow the Special Weapons Manual Lock Handle. Multi-tasking was not a common word then but it should have been.

When I got back to flying status and I tried once again to explain to her about the B-52's receivers, jammers, chaff, and flares, and how they defended the plane and the rest of the crew from the enemy - my mother was lost again. Trying to explain flying low level at tree-top and sitting on alert with nuclear weapons ready to go to war didn't seem like something a mother needed to hear either.

Finally I told her I talked on the radio when we were flying and I assisted the gunner in getting the coffee and inflight lunches for our flights. She understood that, but I don't know what she ever told her friends what her Air Force son did.

Chapter Two

Cold War [kohld] [wawr] – *noun* - A state of political tension and military rivalry between nations that stops short of full-scale war, especially that which existed between the United States and Soviet Union following World War II.

Deterrence
Tom Jones

At Carswell, 1978-83, the 7th Bomb Wing stood nuclear alert with four tankers and eight bombers positioned within the alert ramp, using crews from three squadrons: the 7th ARS, the 9th Bomb Squadron, and my unit - the 20th Bomb Squadron, nicknamed the "Buccaneers."

At each alert force handover, usually on Thursday morning, my crew would join the other oncoming crews for the intelligence and operations briefing at about 0800. Once the briefing ended, we would drive our blue, extended-cab pickup out to our airplane to position our equipment and begin the formal assumption of the sortie from the off-going crew. The RNs and EWs inventoried the coded materials while the AC's talked over the aircraft; the gunners talked shop, too, while the nav and I inspected the bomb bay.

Using a flashlight and a hardware-store inspection mirror, we checked the safety interlocks and settings on our four thermonuclear weapons (whose presence aboard I could neither confirm nor deny to outsiders). Once satisfied the bombs were sleeping peacefully, we aligned the hydraulic actuators to the couplings on the bomb doors, slid home the stubby metal clevis pins, and cotter-pinned them in place.

Using the strong shoulders of the gunners, navs, and copilots, we drove the doors closed with a solid, metallic thunk, taking care not to get sticky hydraulic fluid residue all over our flight suits. Paperwork and inspections done, my crew had the duty. It was now our airplane, cocked and ready.

The seriousness of that weapons inspection has never left me. Every three weeks, I would crawl on a stepladder between two pairs of hydrogen bombs, each capable of destroying a large city along with most of its inhabitants. Any survivors might almost wish they had perished in the megaton-class detonation. The destructive power of our single airplane was almost impossible to contemplate with any sort of nonchalance.

But I also knew that I was ready, along with my colleagues, to execute that assigned mission. We thought it extremely unlikely that we would ever be asked to do the unthinkable, but we also knew that it was our proven training and readiness to perform such a doomsday flight that kept the Soviet Union from ever supposing that a nuclear first strike might prove successful. We, and dozens of crews like us, along with missileers and submariners, made the Soviets' odds of avoiding a catastrophic retaliation so remote that we hoped they would never seriously contemplate an attack.

When SAC's bomber alert commitment ended in 1991, I took inward satisfaction from my small five-year contribution to deterrence. We helped maintain the peace until our major antagonist had collapsed, no longer capable of initiating a major nuclear exchange.

My Very First Alert Tour
Scott C. Barbu

Nuclear alert can be an intimidating thing, especially when you possess only a lone gold bar on your shoulders, which means to everyone else that you are basically clueless and brand new to the Crewdog world. But this implied cluelessness does have its benefits, meaning you can get away with more stupid stuff, especially in the first two or three months you are there.

After completing UNT at Mather in 1986 and EWO training in 1987 and Combat Crew Training three months later, I was off to my first assignment in Guam, U.S.A. I remember taking a five-hour flight to Hawaii and then a seven-hour flight to Guam on a Continental Air Micronesia jet. The second flight had screaming toddlers on both my right and left side the entire trip. I was never so glad to get on the ground. We landed just as the sun was rising, and it was already hot and steamy. I would have to get used to the steamy part. Living on Guam is like walking into a sauna bath and never leaving.

I was assigned to the 60th Bomb Squadron and immediately started filling the squares for mission-qualification. I was assigned to a temporary crew that included Major Mark McGeehan as the AC, and scheduled for several introductory flights. Mark would only be with us a brief time before he went on to "bigger and better" things, but I remember being very impressed with his leadership ability. The typical "local" training mission consisted of a takeoff followed by air refueling, which was then followed by a low-level training mission over two small barren islands in the ocean which we used as targets for both pretend and live ordnance. On return to Guam I would have to jam a simulated echo-band AAA radar, and then it was two hours of takeoffs and landings, known in the scheduling office as "transition". I grew to dislike that phase of flight the most, because there was simply nothing for the gunner or me to do but sit there with our straps and masks on and wait for it to end. At least the nav team had flight instruments.

Now on to the clueless part. I had been assigned to a permanent crew, with a different aircraft commander, and we had just passed the alert certification with the wing commander. The next day we were slated to pull our first alert tour. I remember being slightly nervous the first time I went through that gate – praying I did not screw up the challenge-response or password part. I was told that would get me "jacked up" – meaning the guard would lay me down face first on the asphalt, point his loaded weapon at me, and scream questions in my face. Once we got inside, we attended all our briefings, and we immediately heard a rumor of an upcoming "horn" or klaxon. These horns were our signal to rapidly run out to the aircraft and start engines. They could sound for real at any time, but the practice horns were normally once a week. Our biggest clue on Guam was the fire department. We could see it from the alert facility, and about an hour before each practice horn the big yellow trucks would roll over to the Alpha (alert) ramp to provide fire coverage for any upcoming cartridge engine starts.

So there we were, leaving the briefing, checking into our alert rooms, when we see the fire trucks rolling, not with lights and sirens but real stealthy-like over to the Alpha ramp. Here it comes. Our AC gathered us in a room and went over the procedures: the gunner would drive and park the truck, the EW and gunner would pull the engine covers while everyone else went up into the jet and got ready to go. The last man in gives the signal to start engines and the copilot will then flip the start switch to "cart", igniting the pyrotechnic/pneumatic device on the #3 and 4 engine pod. If all goes well the crew chief should see black smoke coming out briefly from the bottom of the engine pod as the engine spools up.

About 45 minutes later, right on cue, the horn sounded. As brand new Crewdogs, we tried to respond "with extra speed and urgency", which meant our gunner drove so fast he cut off other crews trying to get out of the same parking lot, causing much gnashing of teeth and extended middle fingers. The gunner parked the vehicle right where he was supposed to (out of the engine blast path), and we got out quickly and I took the left wing and the gunner took the right. Typically B-52's had the engine intakes and exhaust covered when parked; they used a plastic red cover on the front and a homemade-truck-inner tube-thing on the back. The gunner was finished with both of his before I got through with my last inner tube, and as I struggled with it, I could see the pilot glaring angrily at me through his side window. I was taking too long, and that was not good. The wing commander was watching. Once I finally worked it free, I threw it in the back of the alert truck with the others and made my way to the hatch, climbed up into the nav compartment. The nav turned around and said "he's in now" and the copilot flipped the cart switch. I then looked down and realized a terrifying fact: I had no idea how to shut the hatch from the inside. I was never the last one in the aircraft, so I had never been taught how to close and lock it. I wasn't on interphone so I couldn't ask anybody, and the noise of the cart start drowned out all attempts at yelling. So I decided to wing it, grabbing the cloth handle and pulling the hatch up while at the same time trying to slam the lock handle forward. It wasn't working. Once, twice, three times I tried. We started to taxi and I decided just to hold it, standing up on either side of the hatch and holding it closed by grabbing the strap while we did our little taxi exercise. I was watching my Nav, and he turned around and looked at me with a genuine expression of "WTF?", then his head turned back to the front and I saw him mouth the words I dreaded, "Roger, simulated takeoff".

The pilot then proceeded to the runway with me still holding the hatch. I still wasn't on interphone. I was sweating hard due to the Guam heat. My unprotected ears were ringing, as I normally had my helmet on during this phase. The pilot advanced the throttles and since I had never done a takeoff standing up before, I noticed that the plane kind of bounced up and down on the shocks as it accelerated due to the uneven runway. The hatch was slipping in my hands and getting harder to hang onto as the plane bounced. Just before I thought I would lose it and go tumbling under the main gear bogies, the pilot reached 70 knots and slowed back down. We taxied to parking, and the moment we stopped, I let the hatch fall open and almost fell out myself due to exhaustion.

As I learned more about the aircraft in the coming months and became less clueless, I often wondered why that pilot had ignored a critical checklist item prior to taking that runway. The "hatches not locked" light on the pilot's side is small but orange and bright. It's tough to ignore, even if the wing commander is watching you. I decided then and there to never let an unsafe situation happen in order to impress those with eagles and stars on their shoulders. If safety is ignored people die, plain and simple. That point would be driven home to me six years later when my first, "temporary" AC, Mark McGeehan, perished in the Fairchild accident, when a "rogue pilot" in the left seat decided to see if a B-52 could execute a break turn in the visual pattern. Mark was in the copilot seat because no one in his unit would fly with this guy, and him, being the squadron commander, stepped up and flew in their place. That squares perfectly with how I remember him, and the kind of guy Mark was.

Cold War TDYs 1975-1987—a Personal View
Ken Schmidt

I spent five years at Carswell AFB in Texas - from April of 1975 to May of 1980. During that timeframe, I only went TDY twice. Once was a Red Flag mission which ended at March AFB, California, with an ensuing flight on a KC-135 for the Nellis AFB debrief. Later I went on a practice dispersal to Bergstrom AFB, Texas. Total time TDY for those two events: five days. Other crews got to do B-52D swap outs to Guam. Other crews did tests and evaluations on systems. Others participated in air shows at distant bases. But for me, the chance for TDYs was not an option. That changed quickly after my reassignment to Minot AFB, North Dakota.

After notification of a PCS to Minot, I had to go TDY to Castle AFB, California for H-model difference training. I think that was a six-week program and I took my family with me for a chance to see another part of the country. Following that TDY, we continued the move to Minot. Once at Minot, we finally got settled in at base housing and I completed the squadron check-in and eventually started the flying training part of my checkout as a B-52H radar navigator. I wasn't even checked out when I was immediately sent TDY again. Minot had just been assigned new taskings as part of the Strategic Projection Force and was being sent on a TDY to Whiteman AFB, Missouri, for a "bare base operation" concept. That meant the entire wing (excepting the alert force) would take the remaining aircraft, crews, maintenance personnel and equipment lock, stock, and barrel to operate from an operating location with minimal outside support.

I flew down to Whiteman as a spare/safety observer and got some seat time with an instructor. Most crews flew one mission, and then back to Whiteman for a second mission that would return to Minot. I got to fly with a stan-eval crew, again getting some seat time and flying as a safety observer on the night terrain avoidance (TA) mission. I got a firsthand look at the B-52H night TA capabilities and never felt safer in an airplane. Seeing how a well trained crew, combined with the new technologies (at least to an old D-model guy) of low light level TV and infrared technology of the FLIR and STV of the B-52H, could perform its mission really opened by eyes. (That was a four-day TDY for me.)

I completed the checkout program and check ride in mid-October and then went through EWO certification and also ORI/Buy None certification for the current year. As luck would have it, a Buy None hit and I got tagged to fly as a substitute radar navigator on an outstanding crew. The flight turned out to be one of the smoothest ORI/Buy None's that I had ever flown. In fact the "new guy RN" turned in the best bomb scores for that Buy None, a triple sync bomb run with scores of 115, 118, 120 feet and a final sync bomb release of 200 feet. As a reward for that I was placed on a crew that was selected to go TDY to Guam. If you have access to *"We Were Crewdogs II "*, see my story on "SAC Rewards Those Who Serve". Anyway, my second TDY of the year was for eight weeks at Andersen AFB, Guam. During that time, I flew three missions: one a familiarization sortie, the second as an augmentee on another B-52 on a 28-hour (yeah, that was a 28- hour) sea surveillance mission to the Indian Ocean, and finally a 14 hour mission. We air aborted seven hours into our own 24+ hour sea surveillance mission. That TDY occurred from December 23, 1980, through mid February, 1981. We got back to Minot just in time for a practice deployment mission. Figures.

As the spring time approached, I was asked if I would be interested in going on another TDY, this time to England as a participant in the last to be held British Bombing Competition, Giant Strike XI. It involved Vulcan bomber units of the RAF and four SAC B-52H crews - two from the 319th BW from Grand Forks AFB, North Dakota, one from the 410th BW from KI Sawyer AFB, Michigan, and my crew from the 5th BW from Minot. A fifth B-52 H (spare aircraft) was flown over by members of the 1st CEVG from Barksdale AFB, Louisiana. In preparation for the 5th BW participation in Giant Strike XI, a special crew was formed consisting of Capt Tim Kaufman, pilot; Capt Don Olynick, Copilot; Capt Ken Schmidt, radar navigator; Capt Dan Romer, navigator; Capt Gene Spencer, electronic warfare officer; and SSgt Bill Rabenold, gunner. Our crew spent two months of intensive training in low level navigation and bombing before departing for KI Sawyer. After spending a couple of days at KI, all five B-52s took off and flew in cell formation to England.

Our base of operations was RAF Marham, a Victor tanker base. All thr crews were billeted at the RAF Open Mess; I think that we were placed in the condemned section of the Mess as the facilities had not recently been used. Specifically, the showers would only produce a thin stream of rusty water, until one of the CEVG guys did some "maintenance" on it, and the bathtubs had large rust stains in them. Anyway, it was made livable for us. Once in England, we flew approximately six training sorties in preparation for the bombing competitions. Of course, we did have some quality time off and got to see the local sites as well as go to London a couple of times. We also got to partake of some of the excellent pubs as well as the local food and drink. However, I digress here. The actual competition was a disaster with all the B-52 crews placing at the bottom of the rankings. Our crew was the luckiest of the SAC guys though since our plane had some serious bombing and navigation system problems (black boxes, cans, amplifiers, etc) right after takeoff. The problems could not be repaired in-flight that basically had us go through the route pilot visual. Our crew was spared the wrath of Vice-CinCSAC who flew over for the awards presentation. One of the Grand Forks Crews had to stay an extra week and refly the route. Not one of SACs best days.

Back at Minot, our crew was disbanded and I was sent TDY again after being home for only two weeks. That time it was back to Castle AFB for instructor training at the CFIC detachment. It was a six to eight week TDY which consisted of academic and flight training. Back to Minot for more simulators, flights and then a check ride as an instructor.

Things were fairly quiet for the next few months (you know, the usual, alert, flying, ORI/BUY NONE, alert, flying, etc.) until April when it was time to go TDY again. The 5th Bomb Wing was tasked to deploy again en masse to Biggs Army Airfield at El Paso, Texas, for another Strategic Projection Force Exercise known as Busy Prairie in which the entire unit, B-52s, KC-135s, aircrews, support personnel, maintenance personnel and equipment, etc. deployed for five days. During that timeframe, we would fly a mission, recover at Biggs and then fly a mission recovering back at Minot. At least it was a chance to get away from the grind at Minot. (A side note here—this deployment happened at the outbreak of the Falklands conflict.)

Soon after my return to Minot I was given a chance to get off the crew force and got a job on the wing staff as a bomber scheduling officer. I thought that just maybe, my TDY days were behind me. But not so fast! During the last few months that I was at Minot, the wing was converting over to the new Offensive Avionics System or OAS. It was then when I was notified that I was being reassigned to Castle AFB as an instructor. "Not so fast." I said to the bomber detailer at MPC. "I'm not even checked out in the new system". He told me not to worry, that problem would take care of itself. Not one to let things "take care of themselves" I got in touch with the Castle guys that were TDY to Minot and got myself run through the OAS conversion. Next I got myself an instructor check (much easier to do as a bomber scheduler) and was then "blessed" as an OAS certified instructor. We left beautiful Minot in late February of 1983 with a static outside wind-chill temperature of 65 degrees below zero and waved bye to Minot in the rearview mirror. Our furniture shipment was delayed a few days because the truck's engine froze up.

Once at Castle, I was assigned as an academic aircrew instructor in the OAS program. One of the benefits (and I use that term loosely) was that the instructor cadre got to go TDY to bases undergoing OAS conversions. My turn came just a couple of months later in late June of 1983 when I got tasked to go to Hooterville, better know at the time as Blytheville AFB, Arkansas. Instructors would rotate in and out every three to four months. During that time, we would function as both instructors, simulator instructors, and flight instructors. We would be housed in the base VOQs until the 1st ACCS would recover at Blytheville and then we would have to relocate to a motel downtown. Once my four months were up, it was back to Castle. About 10 months later, my turn came up again to go TDY - back to Andersen AFB, Guam. Same old place as before but with new faces. Same old grind again—instructor, simulator instructor, flight instructor. That time, I got to stay in the VOQs, but shared them with the geckos. After the TDY, it was back to Castle, and to a staff job as the 93rd BW Air Weapons Officer. That stint was not bad at all, but still I had a couple of TDYs - one to the On Scene Commanders Course as part of my additional duty as Wing Exercise Evaluation Team Manager and another to the SAC Air Weapons Officer Course.

Morale to this story: if you aren't happy with no TDYs, be careful what you ask for.

In all actuality, I had it made very well in my Air Force career. I was only assigned to four SAC bases: Carswell, Minot, Castle, and Offutt and only had to move five times. Although I had my share of TDYs, others had it a lot worse. I never had to deploy to war, but did "fight" in the Cold War, the one we "won" in 1992 which saw the end of SAC. We did things right in SAC (maybe not correctly all the time) but in light of what has happened in the years since SACs disbandment, I feel that SAC should be reconstituted. You know the old saying, "If it ain't broke, don't fix it". "They" should not have fixed SAC and TAC, but left them alone.

The ejection seat that Captain Gerald Adler rode to his miracle landing is on display at the Moosehead Riders' Clubhouse.

The Elephant Mountain Incident
Jerry Adler

It was just another routine training mission. But this one would end in headlines which spoke of disaster and search, joy and tragedy. The story of disaster took place in less than a minute, and in fewer than five seconds of that time seven men were doomed to die while only two of us would live to tell the tale.

The airman often defines flying as hours and hours of boredom punctuated by moments of sheer terror. Flying the B-52, the free world's largest and most potent guarantor of the Strategic Air Command's motto 'Peace is Our Profession' is never really boring, however. The multitudes of intricate systems that make this $8,000,000 beauty fly are too complex to allow much relaxation. But with knowledge of the airplane, the safest tactical aircraft that the United

107

States has ever had, comes confidence and assurance that its crew will return safely to fly another day.

This mission on the 24th day of January 1963 would be different. An Instructor Pilot and Navigator from the New Mexico base were aboard to check out two Pilot-Navigator Instructor Teams from Westover AFB, Massachusetts, in a new terrain avoidance low level navigation system. It is no secret that Soviet defensive technology has forced low level navigation and bombing on us. But when something goes wrong at 500 feet or less there is not much time to correct it or get out.

There was some apprehension about this flight. It was to be the Wing's first such flight after previous cancellations due to weather and faulty equipment. The new system and the unusually low altitudes to be flown caused some of us on the crew to wear even warmer clothing and carry more survival gear than normal. Because of the unusual crew composition, this was to be my first ride in an upward ejection seat in four and one-half years of B-52 flying. Normally, the Navigator-Bombardiers occupy downward-firing seats which require at least 400 feet of altitude for successful operation.

Later I came to ask myself why I wore not only my winter flying suit, thermal underwear, and thermal boots, but also at the last minute I had placed a pile cap, extra warm gloves, matches and a flashlight into my pockets. Why also did I take that upward seat, and study its operation so thoroughly before take off? These like subsequent questions must always go unanswered.

It was scheduled to be only a short flight of some five hours duration. For heavy bomber crews used to 24 hour flights this is like a trip to the corner grocery. But it would end some two and a half hours early near a place called Greenville, Maine, in the beautiful but rugged Moosehead Lake region.

Maine is a marvelous vacationland abounding in fish and game. But in mid-winter it is no place in which to be wandering or even worse in which to be abruptly deposited.

On this particular afternoon with the blue sky dotted with occasional fleecy and intermittent light turbulence the crew was only dimly aware of how cold it was outside. As the treetops skimmed by at

300 miles per hour a few hundred feet below, there was little thought of snow depths beneath.

One moment it was blissful, if arduous, training. The next moment came the turbulence as we approached a mountain range on our planned course. To the Air Force, flying safety is paramount in all training situations, and when moderate turbulence is encountered at low level, SAC's instructions are to climb immediately. This is what we did but the turbulence became worse.

Then came that moment of sheer terror.

There was a loud noise from somewhere. The aircraft snapped into a steep right bank, descending at the same time. Nothing that the Pilots could do would bring it out of this attitude. With Westover's most senior standardization instructor pilot at the wheel as crew commander, the airplane was still out of control. Later they found out why.

No pilot can fly the B-52 after its vertical stabilizer (tail fin) snaps off.

Elephant Mountain passed underneath at what appeared to be touching distance. From my seat I could see out a window normally beyond the line of sight. I looked for the signal to go. The abandon light was on. I groped for the handles which would actuate the seat. It seemed an eternity but it could have been only an instant.

Every flyer wonders what it is like to eject. He hopes he'll never find out. It is not surprising, therefore, that as the seat fired through the top of the fuselage that I thought was 'so this is what it's like to eject'. There was no fear of dying but just the answer to that old curiosity.

I remember tumbling twice in the seat and held on for dear life to what was my only link with the airplane - an airplane that was already a mere pile of rubble identifiable only by its tires.

I don't remember coming through the trees nor can I recall landing. When I came to after an undetermined period of time the quiet was overwhelming. Ahead of me I could see smoke from what I presumed to be the airplane. And around me were the trees, the snow, and stillness.

And I looked at all this from my throne in the snow - my ejection seat. Now this is theoretically not possible. The book says that when an upward firing seat reaches the top of its arc, man and seat should separate, the parachute will automatically open, and from then on a normal landing is to be expected.

Yet it did not bother me that to still be in the seat was not normal. Nor did the fact that my main parachute was still intact, unopened, in my back pack concern me. As I looked around I could see that the small white pilot 'chute had popped out of the pack but this could only have occurred at impact. After all, I had hit the ground with an estimated force 16 times the normal pull of gravity.

All that I thought about was that I appeared to be intact, that they would come after me shortly, and that I was probably the only one to get out, and I was mad at the wilderness and snow, not even thinking that it was these trees and deep snow that saved my life. I was just mad that there was no place to go.

However, I wasn't the only one to get out. Within those last five seconds the other occupants of the upward ejection seats had also left. The two downward seats probably never fired. The gunner and other crew members without ejection seats had no time to manually bail out.

My ejection altitude was probably about 100 feet above the ground but only 20 to 40 feet above the trees. The pilots leaving split seconds later had slightly more altitude as the mountain dropped away from us faster than the aircraft was descending.

But at our speed and angles of dive and bank, the experts say that seat separation and parachute deployment cannot occur. Yet it did occur - for the pilots. The 40 knot wind blowing at the time was probably the big factor in initial 'chute blossoming. The surviving pilot said that the concussion from the aircraft's impact and explosion finished the inflation.

Both were blown away from the flames. But the copilot was killed when the wind swung him into a tree. The pilot was carried over the mountain top and back along the line of flight. Although still carrying the scars of his ejection injuries, he later returned to the cockpit.

Thus on the other side of the line that we all walk, seven men died in an instant of time. The dreams, hopes, and aspirations of seven families were wiped out. Six women were widowed and nine children orphaned by a routine training flight in this fight to keep the peace that is not quite peace.

My broken watch said 2:52 on that winter afternoon. From then until about 10:30 the next morning was a time for waiting and surviving. They later said that the temperature went from 10 below zero at the time of the crash to 35 below. The snow was some five feet deep on the level with drifts to 20 feet. But this itself would not have been too bad.

Despite my injuries, which were later diagnosed as a fractured skull and three fractured ribs, we were trained and equipped for extreme conditions. But when I worked out of my parachute harness, there lay my survival kit imprisoned in a vise in the bent seat. So near and yet so far was the sleeping bag it contained.

That seat, with what I must believe to have been a "Heavenly Copilot" had brought me through the trees taking off several thick branches en route and landed me upright in the snow. Perhaps that inaccessible survival kit was my fare for the ride. There wasn't a scratch on me from the trees. Had I landed any way but upright I could not be writing this story.

The orange parachute then became my means of providing a marker for searchers as well as forming a bed and shelter throughout the night. As dusk approached and no searching aircraft came in sight, a strange doubt overtook me. The smoke that I had seen was no longer visible. I feared that I had panicked and left a flyable airplane. I didn't know whether upon return I would be welcomed or court-martialed.

Little did I know of the massive search operation that was taking place.

Rescue helicopters and other planes from Labrador, New Hampshire, and Massachusetts were brought into play aiding the search parties from Maine based Air Force Units, the Civil Air Patrol, the Maine State Police and Game Department plus many other volunteer civilians.

Some 80 men comprising a ground search party were deploying into the area.

As the evening wore on I thought more often about my family and the anxious hours that I knew would be theirs. I knew that I was all right, but they didn't know that. I thought about our many discussions centering about an Air Force career. Flying to me was not a dangerous profession. I felt that the hazard pay we received was remuneration for a grueling, demanding job in the air and for the 70-hour work weeks to which SAC limited us. Accidents always happened to the other guy, always, that is, except this time.

During that long Maine winter night I passed out for what must have been hours. I tried to stay awake by thinking about my wife and two little girls, telling myself that I would get back to them fearing to go to sleep. But the shock and injuries proved to be too much.

When I came to, I found one boot in the snow, cold and wet. The other was cold although not in the snow. I tried to stand up to attempt once more to light a fire in the wet snow-covered branches. But I couldn't stand. The frostbite later took its toll of one leg and the toes on the other foot.

Finally the planes started coming during the night. I knew that it would only be a matter of time now. I was sure that a searching tanker plane had spotted my flashlight's SOS. Weeks later I found out that it had not, Nevertheless, the thought buoyed my hopes during the night.

With full daylight came more search aircraft. A Maine Game Warden's plane first spotted me. By now I was only strong enough to wave one arm as I lay wrapped in the folds of my parachute. A helicopter from Otis AFB, Massachusetts, hovered at treetop level in what I am told was exceedingly difficult maneuver and dropped a man down to me in a sling.

That paramedic was the most welcome sights I have ever seen. I'll never forget his incredible handlebar mustache. This was my first helicopter ride and I couldn't have enjoyed it more.

I later became unconscious for five days with double pneumonia. That was the worst time of all. I guess the doctors at Eastern Maine General Hospital in Bangor did all they could and then said that it was out of their hand. But God stayed with me.

After an experience of this sort, the unanswerable questions remain. Why me? Why was I tapped on the shoulder? Why was my ejection successful despite an un-deployed parachute? Why was I in the upward seat at that particular moment for the first time in all my flying?

So many imponderables remain. Each separately can be answered logically. But can logic explain the combination of all the factors?

Eventually the questions became dimmer in the mind. I am reminded, however, every morning when I strap on my new leg. Perhaps we should all be reminded more often of that very thin line that we all tread and of the right way to live while we are on this side of it.

113

Oil Burner
Nick Maier

The *"We Were Crewdogs"* books have covered the spectrum of flying the BUFF. There is one critical experience that has not surfaced to remind everyone of the fun and games during low level flying. This story is also for those who have not strapped into a B-52 cruising at 400mph, 500 feet above the contours of high desert mountains, chasing jackrabbits and herding wild mustang horses through Monument Valley.

SAC bomber crews normally flew a minimum of three Combat Crew Training Missions (CCTM) each month, all of which simulated an Emergency War Order strike profile. Except for the timing, routing, and targeting, the CCTMs rarely varied in content. The max heavyweight takeoff, air refueling, and two-hour high level navigation runs would be the precursor to the grand finale, an Oil Burner challenge. The Oil Burner designator for low-level training routes was appropriate, considering how the B-52's smoky engine exhausts were a well-known signature. We also anticipated each contour-following flight mission for its uncompromising demands.

The most difficult task was to get our minds in gear for the early morning flights we had dubbed, "the dawn patrol." They usually started the same for all bomber Crewdogs. A screaming wind rattled the walls of base housing, signaling frontal passage of another blue-norther. The body could never get accustomed to Oh-dark thirty get ups. Stumbling around in a daze getting showered, shaved, and suited up had become a ritualized rote for SAC flyers. The next primary hurdle was accompanied by a silent prayer that the family wheels still had sufficient battery juice remaining to start in the sub-zero cold. Then we audibly cursed, nursing the struggling car through four inches of new snow driving to Base Operations. These traumatic events were harbingers to the two-hour physiological battle against debilitating wind chills it would require, to get our aging B-52D safely airborne one more time. I often paraphrased an old brown shoe Air Force adage to "kick ten frozen tires and light eight J-57 fires" to get our cold soaked bird moving down the taxiways.

After the sweaty palm two-mile chase down the runway, a heartbeat behind our buddy tanker in the pre-dawn takeoff, it still took a full half-hour to warm the cold soaked crew compartment. It was no problem following the KC-135's greasy black engine exhaust trail, as it climbed up through the gray overcast murk that we had to live with eight months out of a year in the heart of America's Siberia. Poking holes through the mournful light of a winter morning, the mated tanker and bomber leveled off at 29,000 feet to enter the box of airspace reserved for our in-flight refueling duet, and traded 30,000 pounds of JP-4 jet fuel. The total tanker and bomber connect time was less than 10 minutes. We then faced a three-hour high altitude navigation profile, which approximated the long distance flight approaching an adversary's early warning radar net. The realism of flying over endless stretches of featureless terrain satisfied another aviator's cliché; "Flying consists of long periods of absolute boredom, punctuated by moments of sheer panic." The radio finally crackled to life and jump started the crew for a unique experience found only in SAC.

"Borax-29 is cleared to Oil Burner Seven; report when departing flight level 37. This is Thermal radio over!"

"Roger, 29 leaving 37-thousand, estimating exit at one eight four five Zulu." The clipped voice of my Zoomie Academy copilot was distinctly military. He had that detached monotone of a SAC bomber crewmember that had transmitted this radio call innumerable times.

"Pilot, start your descent to 25-hundred feet and take up a heading of three two zero. This'll put us at a 500-feet terrain clearance setting, and on course to the Hodge turning point, estimating Hodge at 1720 Zulu."

"Roger Nav, descending to two five hundred, three two zero, and seventeen twenty. We're going to scare the hell out of lot of jackrabbits today."

"I'm not worried about the bunnies Ace," the RN replied. "It's the bloody doctors and lawyers flying around down there with their heads up and locked that can ruin our whole day."

We descended through a thick layer of "mare's tail" cirrus clouds at 35,000 feet. When we were flying over them straight and level, they appeared to be transparent. The bomber was through them in an instant, dropping over 4,000 feet a minute. The noise level in the crew compartment increased geometrically to the reduction in altitude and acceleration of the aircraft. Naturally, so did the level of turbulence. I would be muttering out loud as I wrestled with the flight controls, "It won't be long before the stadium wide wings of this beast I'm flying will be flapping over Mother Nature's vertical speed bumps." In the distance, I could already see the first shimmering thermals rising up from the increasing mid-morning heat. My control wheel transmitted the first of many wind sheer microbursts, as I dropped the BUFF into the swirling tubes of unstable air seeking equilibrium.

The crew's mission was to navigate a low-level training route that would take them from the entry point over Thermal, California, to a simulated nuclear weapon release on a target near Mono Lake, just east of the Yosemite National Park Tioga Pass exit. It would require flying at a ground hugging 500 feet above the Mojave Desert floor, centered on a microscopic radar beam through the undulating mountain contours alongside US Highway 395. The weather was clear and unlimited visibility, which accentuated every boulder, and provided a scenic airborne tour of the California high country. I noticed the high wispy clouds on the northern horizon, which were an unwelcome sign of the frontal weather that was briefed prior to takeoff. If our luck held, we should pass over the target and be back at cruising altitude ahead of any thunderstorm buildups. Before this day ended the crew would be strapped into our bomber almost 10 hours. The entire mission was a humdrum, routine trip for all SAC aircrews, who trained for the day deterrence and stalemate might fail.

My bombing team was as dedicated as any in the Strategic Air Command. They were also not immune to a mysterious malady, which infected combat crewmember's state of mind on low-level missions. For reasons known only by the participants, the penetration from cruising altitude to the closeness of terra firma somehow altered their psyche. Perhaps it was the arid air of the vast wasteland in western America, where 90 percent of the Oil Burner routes were established.

The rationale for these route locations was the similarity of ground features expected to be over-flown during a retaliatory response against the Soviets. The American farmer-rancher folk were also sufficiently scattered about to reduce noise and air pollution complaints to an acceptable number. The aircrew's repartee would be a psychological buffer between the reality of what they were training to do, and the terrible consequences of what they would be required to accomplish if they ever had to fly a strike mission.

Grunts in trenches, marines in assault ships, and swabbies in steerage battle stations, all survived because of the insane humor of their predicament. For my crew, the first symptom of their altered egos could be attributed to the rapid change in cabin pressure, which occurred during the maximum descent designed to outwit the enemy's ring of long range radar defenses. SACs great iron birds would plunge unseen with their wing spoilers completely raised for full drag effect, their power plants at idle speed, and their air conditioning systems in disarray. The crew compartment dust and aromatic debris was finally released from suspension. The men strapped inside switched mental gears, and in the process became an entity worthy of intense study.

Both pilots shared equal responsibility for several critical tasks, designed to get the weapon on the target in one piece. Sitting shoulder to shoulder manhandling their control yokes, rudders, and throttles to remain somewhere within the briefed flight envelope; they cross-checked the navigator's heading and altitude requests using visual confirmations. Reading road signs and mile markers on the deck at over 400 miles per hour worked well if everything was written in English. Contour flying across the moonless steppes of Siberia, or in the scud of weather, would require following along meticulously with compass headings and time ticks marked on their own low-level route map.

This four-color cartography was a very useful tool of a bomber pilot's trade, if the nimble fingered airplane driver hadn't spilled too much coffee on his work of art. The accuracy of this dead reckoning procedure is totally dependent upon the assumption that the pilots and the navigator started their mariner's nightmare at the exact pre-planned point in space, and at precisely the correct instant of time. Under Instrument Flight Conditions, this wrinkled; sweat stained touring map would be as functional as utilizing a half-filled cup of coffee for an aircraft attitude indicator. SAC bomber crews also looked upon low-level flight with their characteristic morbid humor. A more realistic definition of the acronym SAC was, "scattered around the countryside. At the moment my copilot and I were totally oblivious to the surrounding high decibel roar of teeth-jarring turbulent air stream, conducting an unending crosscheck of aircraft systems indicators.

"Pilot, I have you approaching level off at 25 hundred feet, maintain course three two zero."

"Roger, Nav. 25 hundred, three two zero, plus or minus 10 percent on this washboard. This is a rocky road already, and we're not even in the target area yet. "

No one bothered to comment since they were all enduring their own private hell trying to survive the physical and mental anguish of the increasing turbulence. Every third or fourth cycle measured about eight on the Richter scale, which occasionally redistributed the dust and loose debris in the crew compartment. There was no way that the autopilot was going to hold the altitude, so I would get a workout in upper body fitness training.

The control wheel gyrated in a rout pattern that was maddening to follow, causing my forearm muscles to react to the strain immediately. The copilot had his left hand manipulating the eight engine throttles, trying to keep the airspeed within a 10-knot ballpark. Any attempt to hang on to his control wheel with his right hand was abandoned. His knuckles had been rapped too many times in the past. He would take his chances when he was forced to grab his mike switch to trigger the radio closer into the target.

The aircraft's constantly shaking instrument panels reduced the incredible array of gauges to an unreadable blur. Through the glaze of vibration, we somehow managed to freeze-frame interpolations and successfully monitor our flight status: fuel consumption, - 15 gallons to the mile; 80 aircraft systems gauges, - a single abnormal needle position could cause mission abort; 99 emergency warning lights, - one red flash demanded instant instinctive pilot reaction; and 26 flight instruments, - the trick was to keep them all aligned to the conventional attitude for normal flight, preferably right side up.

Staring at each of us through this visual cacophony, were our individual terrain clearance radarscopes. Even though we had daylight visibility today, it was important that we constantly reinforce our skills flying the radar generated horizon indicators. The next mission could be in the weather or at night, forcing us to discipline our absolute reliance on the electronic contour following trace line. The eerie traces would outline any hard centered clouds ahead, directing pilots to gradually climb-to-clear. Failure to comply automatically resulted in six grieving families, an accident investigation board, and the possibility of having their name over the gate of an Air Force base.

To add to the intricacy of this ground hugging flight profile during stateside training missions, all crew members listened intently on two radio frequencies for any signs of life from Air Traffic Control, the Radar Bomb Scoring site, or anyone else who knew or cared that they are down there among the sand dunes. Somewhere along the way, each man on board also faced the supreme challenge of peeling a half-cooked hard boiled egg, to enhance the odds that they could keep down their gourmet, in-flight kitchen sandwich of baloney and suspicious smelling cheese.

The turbulence on the route was multiplying geometrically. The noise level in the crew compartment battered the senses as much as the visual activity that was required to keep the entire show going. Everyone on board would be hearing the rush of air and dull drone of engines for many hours after the mission. My eyeballs had long ago established their pinball machine rapid scan over the dozens of instruments required to keep my aluminum overcast safely airborne. I was never surprised to notice that my pilot friends had the identical "track while scan" look in their eyes.

I believed that if all of the instrument needles were pointing in approximately the same direction, somewhere between 12 and two o'clock, the airplane would counter the effects of gravity and whiz about the sky without blowing up. The panic began when the pilot's eyes registered an erratic bouncing, slowly decaying, or bent around the peg zero indication that signaled disaster. The red emergency lights that usually accompanied those hellish round dials were equally demanding, requiring an immediate instinctive response by the now cursing airplane driver. Red was truly the universal beacon for emergency notification. Unfortunately, it also matched the color of the cockpit when it was engulfed in a giant ball of flaming aircraft fuel. The trick was to do something; take some action, push a button, pull a fire shutoff switch, scream at the copilot, or whatever had been cleverly designed by the creative aero-engineers that would extinguish the blasted warning light. The brain had to be quick, the memory accurate, the hand steady, and the voice collectively cool, while the pilot waited that infinitesimal heartbeat of time for the light to blink out and remain inert. He would shout obscenities, and make silent prayerful promises to his maker, if the stubborn warning persisted. Depending upon the outcome, he would either have another war story to tell his grandchildren, or they would always debate the probable cause of that fatal crash dimmed by memory.

It was usually at this point in our sight-seeing adventure that the Radar Navigator interrupted my concentration with an impatient query, "Will someone upstairs check and see if the pilot is comatose? My scope just blanked on me, and the nav's is strobing like some kind of jamming hit us. Pilot, climb 500 feet for radar inop terrain clearance. I'm going off interphone to do some combat boot maintenance," the RN said in his usual mile a minute ticked off tone of voice. Five feet aft of the pilot's cockpit and six feet below, down four metal ladder steps, in a dark, dank, hole carved out of recycled metal, sat two ashen faced Crewdogs who had been training for eons to navigate airplanes and drop bombs. Facing forward on the left, the RN maintained a watchful right eye on his younger protégé, fresh out of "East is least and West is best" navigator training. The nav tried desperately to convince the rest of the crew, that he knew where he was on the globe's surface within 100 miles at any given moment. The old pro RN would grab his bombing control handle; expertly zig and zag the aiming crosshair trace across the radar screen to an identifiable ground return, yawn, and inform the nav that he couldn't navigate himself to the latrine.

Life was definitely not a rose garden in the nav team's hellish cavern. The whirring and clattering of electronic gear was acoustically added to the preposterous in-flight noise level. The screaming, throbbing heartbeat of air driven alternators and air conditioning equipment was located in the forward wheel well, a scant four feet behind the thin walled bulkhead of the pressurized crew compartment. This din resulted in a gradual deterioration of their hearing, and competed with a secondary amenity that also came with the territory. A horrific odor which would defy chemical analysis.

By virtue of this sinkhole's physical location, and the strategic placement of the forward crew compartment's air conditioning flow regulator, the polluted stench of stale coffee, sour milk, countless cigarettes, periodic vomit, five to six different ethnic sources of digestive gases, and every germ carried aboard the aircraft, percolated into an invisible toxic cloud. This deadly, swirling cyclone drifted downward and aft to engulf these unfortunate aviators, along with the dust and litter generated by endless hours of flight. The epitome of insult to the side-by-side occupants of this low brow real estate was the masterful selection of their neighborhood for the location of the crew urinal and commode. Their prominent positions in the main traffic area during any circulation of crewmembers placed them as testimonials to the one factor limiting the range of this huge intercontinental bomber - crew endurance. The lower compartment's total biochemical combination of odoriferous atmosphere was a perfect mask for the microwave zapping each crewmember was enduring from the aircraft's radar magnetron. This silent generator emitted its gene bending energy from an unobtrusive position only four feet forward of the unsuspecting lower deck residents, and almost directly beneath the two pilots. The long-term affects of the radiation exposure would someday become evident, when the aircrew's future progeny suddenly emerged malformed or maladjusted.

In response to the RN's radar malfunction, I was out of my seat and parachute harness in a practiced instant, and executed a half roll to the right and aft toward the lower deck ladder. In spite of the bloody turbulence, the RN sounded like he needed some help. I knew that it was very rare for him to leave the protection of his ejection seat while on a low-level route. He could usually do most of his radar repairs from his seat, so this must be a major malfunction. By the time I shimmied down the four steps into the confines of the Bomb-Nav stations, the RN was sweating and cursing at the electronic innards behind a panel that he had already removed from the brains of his computer system.

"I'll keep this in pilot level terms, Ace," he said loud enough for me to hear above the tremendous noise level surrounding us. There was no decorative airline-soundproofing cocoon on a B-52, just bare bulkheads and peeling decals that used to be informative.

"This whatchamacallit is disconnected from that thingamabob over there, you see? Hold on to this black box, you won't get too much of an electrical shock if it's working right," he yelled in a smiling bark. "Once I tighten these two widgets together, we'll be uptight and outa' sight."

I could see that I was not a critical entity for the situation, so I quickly made a pit stop and headed upstairs. This rocky road had turned ugly, and I banged my knees twice climbing the ladder. The navigator attempted to hide his barf bag from me, but the stench was unmistakable. The entire crew could always look forward to the nav's airsick stories after every low-level run, no matter how smooth it had been. I realized that I would be joining this sick call if I didn't get back into my seat and at least orientate my queasy guts to a horizon. As soon as I was secured into my harness and quickly updated the situation, I allowed myself to relax with some inner thoughts.

"This afternoon's view from my window would make any navigator happy that he was flying sightless. I have either been flying these low-level runs too often, or do not possess enough sense to be frightened. Normal men are content to pursue the mundane aspects of living. Here I sit with 200 tons of airplane in my grip, carrying over half its weight in highly volatile JP-4 jet fuel, buzzing the jackrabbits at 400 miles an hour. I have purposely aimed my eight-engine monster to within 2.7 times its own wing span over the hard desert floor, forcing the machine through a washboard pattern of unpredictable turbulence that is patiently waiting for me to make one small mistake.

"It doesn't matter what error in judgment or flinch of muscular coordination I am guilty of - gravity is unforgiving. The last thing I will see is my windshield full of boulders and scrub trees. The last thing I will hear is my incinerated brain screaming at my soul for being careless. The SAC accident board will spend six months trying to deduce the probable cause of losing this $8,000,000 strategic bomber before they publish their catch-all decision: "pilot error."

"Whenever you tourists find the time, I would like a heading correction to three four five, the crosswinds picking up out here," the navigator broke into my soliloquy, sounding a little testy. I replied immediately to confirm the nav's new heading. I also knew that conditions below had deteriorated. By now the RN would also have his face into a completely full barf bag. He usually outlasted the navigator, except under the current conditions. The constant turbulence was aggravated by both pilot's vain efforts to hold heading and altitude within five degrees and a hundred or so feet. That was no small ballpark to stay in, considering the diabolical forces of nature that were continually trying to wrench the aircraft from my hands, and dash it upon the nearest piece of sod less than 500 feet straight down.

The real sense of speed on a low-level run was when the pilot looked out his side window at about 10 o'clock low. The ground rushed by in a blur of undulating fast forward filmstrips. That's why pilots never look down. The sights and sounds experienced from the cockpit would overwhelm an ordinary soul. I was now looking at a sign of dread for all low-level aviators. The weatherman missed his forecast again. Those high wispy clouds that created such a beautiful post card effect less than 20 minutes ago had suddenly turned ugly. The entire mass of air that the crew was suspended in had snapped into the familiar whiteout of haze and lowering visibility that foreboded frontal passage.

Checking my clock confirmed that were still 30 very long minutes before reaching the target. The exit route would force us to climb straight north into the teeth of the storm clouds. I immediately rationalized the dangerous predicament with the standard pilot's axiom. "You pays your money, you takes your chances."

"I'll take the airplane, copilot. Lets get this show on the road, how's it goin' down below radar?" I asked, trying to mentally accelerate events.

"Maybe your coming down here and checking the plumbing helped, Ace. Or it could have been the size of the hammer I used. The scope blanks out intermittently; I just hope I can catch the target in between the blinks."

"We turned over the Keller pre-IP during your morning stroll, pilot. Estimating the IP on time at 1520 Zulu."

"Roger Nav, you did a good job of bracketing the copilot's 10 degree heading changes due to this terrible-ness."

My eyeballs instinctively included scanning the horizon through a bug-spattered windshield during the continual instrument cross-checking. It was up to my memory cells to register that visual data and decide whether there were any impending disasters out there. This gathering of information was impeded by being seated in a surrounding barrier of noise, which was varied only by the range of intensity. Technically I shouldn't have been worried about a midair collision with other air vehicles. The low-level Oil Burner route was off limits to all other traffic, whenever FAA cleared SAC crews to operate within it.

The primary hazard was migrating birds with no radios. A bird strike could result in damage comparable to an anti-aircraft missile. The Oil Burner routes criss-crossed migratory bird flyways, with B-52s competing for airspace with wildfowl of every species. Occasionally, the bomber pilots would see the ultimate horror. There it was - a spot that suddenly showed movement, then growth, and finally recognition. 12 o'clock, level, and closing at a measured velocity of 530 miles per hour.

A sporty red and white private plane carrying some nouveau riche doctor and his sightseeing toothy family was heading toward another unreported "near miss" mid-air collision. Those fair weather pilots flew their shiny new airplanes the same way they drove their Mercedes. They took their half of the highway out of the middle, and flew the straightest line between two points without checking their navigation charts for the bright hashed marked warnings along the Oil Burners. After all the participants of this episode had screamed Mommy and lost their respective lunches during evasive maneuvers, the remainder of the bomber crew without windows, very calmly screamed over the interphone,

"What the hell are you jokers doing up front, practicing aerobatics down here at boulder level, for crissakes?"

The shuddering, recovering bomber droned on. It had accumulated a few more aches and pains in popping and banging its wrinkled aluminum skin, which forced the microscopic wing cracks into a new direction around the stopgap drill holes. The over stretched contorted airframe returned ever so slightly slower to its originally intended skeletal shape, being impeded by the buildup of minute crystals caused by the fatigue of metal being repeatedly vibrated and twisted, exhausting every molecule to the brink of total breakdown and catastrophic failure.

The RN finally came up for air and barked over the interphone, "Whenever you jokers can find the time to get your act together, we're at the IP inbound to the target. Let me know if you get visual upstairs."

"Crew, I've got a search radar at 12 o'clock," the voice of our young EW broke into the RN's question.

"E-dub, where else would a search radar be out here? At three-thirty?" the nav managed to get his two cents in while reading the label in his airsickness bag.

"Bomb plot, this is Borax-29, IP inbound." The copilot's initial call to the Radar Bomb Scoring site brought the crew back to the reality of why they were on this excursion.

"Roger 29, bomb plot standing by for a 1845 Zulu release time." The voice that responded out of the surrounding desert was thin and wasted, which matched the "bored out of the skull" mentality of its source.

The luck of the draw had placed a huddled group of airmen inside a converted railroad boxcar, operating a Radar Bomb Scoring site. Their car full of sophisticated gadgetry had been trundled out to a windblown siding by some passing freight train in the night, and left rusting to the tracks for months on end. The RBS technicians cranked and tuned their black boxes and radar dish, to track and score the theoretical point of impact for the bomber crew's electronic thermonuclear weapon.

When all of the B-52s black boxes accomplished the preset magic the designers intended there were no targets anyplace on planet earth that could not be obliterated with unerring accuracy. The real bomb on the actual "day" would be released by a computer signal, then sway in a parachuted retardation to allow the aircrew some measure of escaping the blast and radiation. After a predictable time of fall, the olive drab monolith gently plants its crushed nose into the ground. The enemy would rush up to see this upright monolith, hear the metallic ticking deep within its dull olive drab skin, and then read the large decal printed in their native Cyrillic.

"Greetings, Tovarich. You have one minute to run, and one minute to dig!"

The RN retched one final time, before the seriousness of his predicament on this bomb run calmed his heaving guts.

"As I was saying pilot, let me know if you ever get visual upstairs. These radar offset aiming points aren't checking out too purely." In between barfs, he was required to switch his wizardly radar equipment from one aiming point to another, both of which were geographically offset in a measured distance, at specific compass directions, from the actual aiming point of the intended target. This tricky little procedure became an absolute must, when it was discovered that those who went about burying ICBMs in underground silos, usually took great pains to insure that nothing remained above ground to produce a nice shiny radar reflective aiming point.

Hence, the bombardiers of Armageddon were forced to navigate their nuclear battle wagons to the weapon release line, using the only available sources of radar returns within range of their antiquated equipment, such as; mountain peaks, oil storage tank farms, the county courthouse, or that bend in the river exactly 5.72 miles at 061 degrees from the target. It didn't take a doctorate in geometry to calculate the effects on accuracy if the aircraft was one tenth of a mile off distance, and/or one half degree outside of the desired bearing at the time of bomb release. All the calculations of Archimedes would not change the known ballistics of a free falling bomb, and place it anywhere near its intended target. Better to have driven it there by the Arkansas National Guard on a flatbed truck.

The navigator saw it first. "Hey, Radar there's the aiming point, it's finally showing! See it? It's shaped just like your crooked schnoz! 60 seconds to go, pilot."

The RN saw no humor at all in the punk faced navigator's remark, and stuffed the remains of his now thoroughly rancid baloney sandwich into the nav's disconnected oxygen mask, which was hiding the kid's twisted grin. At the same time, he slammed his control handle hard left and forward just a tad to move the radar crosshair to the aiming point. With a negative shaking of his head, he reluctantly agreed that this blinking radar return must be the one they had memorized in their long target identification study sessions during ground training.

"Center the PDI, pilot!" he ordered. "That's the black and white steering gauge on your left there, the one with the little white needle!"

"30 seconds to go," blurted the navigator, now retching his own lunch since the baloney had a distinct stench of the RN's vomit. Being a professional upchucker himself, he managed to keep one eye on the time-to-release meter and the other on his rapidly filling helmet bag, which was the closest receptacle he could lay his hands on.

"You gotta be joking Radar?" I yelled over the din of radio and interphone chatter. "The PDI indicates 30 degrees to the left!"

"I didn't ask you how much turn was required to center it, pilot. Just CENTER THE BLOODY PDI, DAMN IT!"

The navigator cut off his partner's tirade by announcing curtly, "20 seconds to go, tone on!"

Without bothering to look, the RN reached to his left and instinctively snapped the Bomb Scoring Tone switch ON, which transmitted a steady, shrill signal over the UHF radio. This raucous alarm startled back into wakefulness the RBS troops, slumbering miles away in their re-upholstered boxcar. They were already on the brink of madness, having been ordered out here to the most remote outposts of any western or northern state, and sentenced to wait and listen for that incessant radio tone.

Tracking the bomber with their own radar system, they were able to score where the imaginary bomb would have hit the ground at the instant the RN's scoring tone finally ceased its shrieking. Their dusty black boxes quickly computed all the factors fed into its electronic brain. Bomber heading, altitude, airspeed at release, known trajectory and ballistics of the weapon's shape, and the assumption that most aircraft delivering bombs were usually flying straight and wings level at bombs away.

I was now faced with an enormous technical problem that an army of aerodynamic engineers could not solve. How could I turn an aircraft the size of half a football field, flying over 500 feet per second, through 30 degrees of arc in less than 20 seconds? Frantically attempting to do the impossible, with my panicky radar's bellowed order still echoing in my headset, I yanked the control wheel hard left.

The jerking crank bloodied the copilot's knuckles, when the distraught apprentice tried to reach for his radio switch and transmit the 20 seconds-to-go warning to the RBS site. At the termination of the endless sounding tone signaling weapon release, the Stratofortress was still in a God-awful 50 degree bank with its left wing tip 97 and a half feet closer to mother earth. The only sound that broke the total silence inside our roaring, struggling aircraft was the RN's mournful, "Daamn!"

"We'll be lucky if this one is even in the right state."

I was startled by the voice, realizing that I had not heard a word from the gunner during the entire low-level route.

"Guns, where the hell have you been during our circus down here today?"

"To tell you the truth, Boss, I was afraid to let go of my handholds back here to talk to anyone. That was the roughest ride I have ever had, my helmet has three new cracks in it from banging against the bulkheads. By the way radar, I saw the target visual when the pilot stood the airplane on its wing during that victory roll. It was supposed to be in Utah, wasn't it?"

"That one liner was a worse bomb than the radar's," I answered.

Both crew compartments were suddenly transformed into a mortuary gloom. The mission pressed on as the bombing team mechanically completed their post release checklist procedures, while waiting for bombing results from the RBS site.

It did not surprise me that there were no comments from the wing staff regarding the low-level gross error bombing score from that miserable mission. The push to prepare for Arc Light operations relegated low-level bomb scores to a "gee whiz" factor. We would all be anxious to pack our bags. SAC was providing us with an escape from another frigid winter by shipping us out to the Pacific tropics at Andersen AFB, Guam. It would be a six-month tour of duty on the Rock, as an Arc Light cadre unit.

He's Coming Home with What?
Ben Barnard

I had just settled into the normal four to five hours of sleep a squadron commander usually gets per night and was in the REM stages when the phone rang. I sneaked a peak at the clock on the nightstand and thought it read 0230. Phone calls at 0230 usually don't have too much good news associated with them and this was no exception. I did have three children in junior and senior high school, so I was hoping that nothing had happened to any of them. They shouldn't have been out at that hour, but I remind you to think back to when you were that age. In addition, our parents were getting up in age. But, it wasn't any of that - it was the Command Post Controller who ranks a close third to concerns behind kids and parents.

I answered the phone the way I usually did, "Lt Col Barnard." The controller responded, "Sir, Capt Swegel is headed back to the base and he has lost two engines."

I said, "OK, who is the duty IP?" somewhat nonchalantly as a B-52 has eight engines and losing two is somewhat odd, but certainly not catastrophic. The next transmission almost made me fall off the bed. The controller then said, "Sir, he's lost two engines - they fell off the airplane." You can probably imagine my response to that - "WHAT?...They fell off the aircraft?"

"Yes sir, they fell off the aircraft."

After a couple of long seconds, I told the controller that I would be there in 10 minutes. When I arrived, Alpha (the wing commander) was already there. The first thing we did was contact Boeing Safety in Wichita and discussed the situation with them. Since that was the first time they had dealt with a situation like that, they could only provide us with what they thought would occur during the landing.

When you are missing parts of the aircraft, you generally don't know exactly how it will react at different airspeeds. Next, we worked the performance charts together and generated some landing data to crosscheck what the crew had calculated. The wing commander then said we had done all we could do in a windowless building and suggested we head out to the runway. After getting into his car, we started out to the hammerhead on the approach end of where the crew would be landing. In the interim, we were talking to the crew via UHF radio.

As the flight manual recommends, the aircrew was going to make a simulated approach at altitude to see how the aircraft reacted at approach and landing airspeeds. They accomplished that after slowly lowering the flaps in small increments due to potential damage to the wings and found all aircraft responses were normal. All the while, we were speaking to them via the radio. By then it was approximately 0400 - still dark, which added a little more suspense to the situation. It was finally time for the crew to attempt to recover the crippled aircraft.

Even though things were tense, it wasn't as if this was the first time they had flown a six-engine approach. The B-52 training manuals require all aircraft commanders to fly simulated six-engine approaches and they are required to demonstrate proficiency on their qualification checks. These practice approaches are flown with one of the outboard sets of engines (either one and two or seven and eight) at idle. The crew had lost three and four so there was less of an asymmetric problem than an outboard pod would have provided.

The pilots did a superb job flying the approach and touchdown was normal. Of course, the full complement of fire department vehicles and ambulances were on the taxiway in order to mitigate any problem which might occur should the aircrew experience problems on landing. After the landing, the wing commander and I barreled down the runway behind the aircraft so that we could monitor the aircraft as it slowed to a stop. After the aircraft stopped, the crew egressed as is directed in the flight manual. With the whole pod gone, we didn't know what would be leaking from the pylon where the engines use to be (fuel, hydraulic fluid, etc.)....and, what kind of fire hazard that would bring. In any event, the crew egressed the aircraft with no problem.

Obviously, the wing commander and I were interested in speaking with the crew, so we went to pick them up. Both of the pilots got into the back seat and when I turned to speak to Capt Swegel, the aircraft commander, the first thing I noticed was his legs—they were shaking uncontrollably. He looked at me and said, "Sir, I can't get my legs to quit shaking." It wasn't funny at the time, but we later had a good laugh about it. We took them back to the bomb squadron and had a nice long discussion about this once-in-a-lifetime event.

What we learned from the crew is even more amazing than the loss of the engine pod. The crew had been flying a night low altitude (at about 400 feet above ground level) simulated bombing mission in northern Maine when they experienced hydraulic failure. That failure probably saved their lives as the crew aborted the route and began a "zoom" to altitude as a result of that problem. It was then that engines three and four disintegrated and exploded and the shock wave from the explosion flamed one and two out. So, at that time, the crew had no operating engines on the left side of the aircraft. That made for a huge asymmetric power problem. Fortunately, the copilot was quick to run the "restart" checklist and they were able to get one and two back.

With that said, things were still tense - the radar navigator said that the vibration from three and four was so intense they couldn't read any of their instruments and things didn't get back to some semblance of normal until the pod departed the aircraft. It was then that the vibrations ceased.

Bottom line: they had been trained to do what they did and they did it near perfectly. The next morning after the maintainers had the aircraft towed back to the ramp and parked; my ops officer and I went out to take a look - very interesting sight to see a bomber which is supposed to have eight engines with only six. The pylon looked normal except for the fact there were no engines attached and there were wires hanging where the engine pod had once been.

The story didn't end there. The investigation revealed that the crew and the maintenance folks had done everything correctly (which is a good thing). The only thing that really caught the investigator's attention was the fact that there was a flight surgeon in the copilot's seat for the flight and landing. In all services, we are identified by our SSANs and what we refer to in the Air Force as Air Force Specialty Codes (AFSC). If you don't speak Air Force, it is known as Military Occupation Specialty (MOS) in the Army. Well, the guy sitting in the right seat of the bomber was coded a flight surgeon via his AFSC, which was true. Pete Mapes, the copilot, was in fact a doc and a pilot - one of six or seven in the Air Force at the time. Once the inspectors discovered that he was a qualified B-52 copilot as well as a flight surgeon, that concern went away. But it was rather interesting while they were ascertaining that fact.

As the coup de grace for the story, the crew was recommended for the Mackay Trophy which recognizes "the most meritorious flight of the year" by an Air Force person, persons, or organization. And they won!

The award was created in 1912 by Clarence H. Mackay, a prominent American industrialist, philanthropist, and aviation enthusiast. The three-foot tall silver trophy rests on a mahogany base and features four winged figures surrounding the cup, each holding a different pusher-type biplane. The winners' names are engraved onto silver shields affixed to the base. The trophy is now on permanent display at the National Air and Space Museum in Washington, D. C. in case you want to see it first hand. They won the award in 1993—the inscription just reads Crew E-21, 668th Bomb Squadron, Griffiss AFB, NY, I believe. However, I feel you should know who these guys were. The crewmembers were as follows: Capt Jeffrey R. Swegel (aircraft commander); Maj Peter B. Mapes (copilot and flight surgeon); Capt Charles W. Patnaude (radar navigator); Lt Glen J. Caneel (navigator); Capt Joseph D. Rosmarin (electronic warfare officer).

The citation reads, "For quick thinking, immediate reaction, and astute situational awareness enabling them to return a crippled B-52 aircraft to stable flight and safe landing." This is on the Mackay trophy website.

As a final note for this story, the pod and engines were never recovered. Somewhere in northern Maine, there is a B-52 engine pod, with engines attached, waiting for some hunter to stumble upon one of these days.

Glasgow Invades Dyess - 1968
Marv Howell

When Glasgow closed, the crews were sent to many different bases. Some went to Castle, some to Fairchild, etc. My crew and a few others were selected to go to Dyess to be instructors and help with conversion to the D-model. It didn't work that way - the AF changed its mind and no D-models went to Abilene. We became the students in the E-model. It wasn't a big deal since we adapted easily. But for some reason, we were viewed as outsiders by the staff. We got along great with the other crews but it just seemed like we couldn't get adopted by the Wing.

The first example of this was seen with our crew hats. Like all bases back then, the bomb wing had its own hat. If I recall correctly it was a red ball cap with the wing designation on the side. Even before we pulled an alert tour we asked about ordering caps. We were told we would have to wait until they decided to do a new order. So what were we to do? My AC was Russ Harmon, a very sharp and feisty captain at the time. He said just wear any hat you want. So we did.

Russ wore the standard Air Force flat hat, The CP wore one of the Tam-O'Shanter hats Glasgow gave us for Arc Light. The Radar chose a Thailand "Go-to-hell" hat and the Nav went with the yellow leather baseball hats we wore at Glasgow in winter. I chose the blue baseball hat with the big dipper stars (also a Glasgow hat). The Gunner wore his wheel-hat. All the others crews got a big laugh out of it, but the Ops Officer was not pleased. We spent about 15 minutes after the briefing getting chewed out. If I remember right, the words were "Harmon you better get these guys under control." Russ was also very cool. He said, "Colonel, we are trying to fit in, but the Hat-Honcho won't order Dyess hats for us." The order was placed very soon after that.

One thing which they had at Dyess was a highly refined CHP (Crew Harassment Program). Even in those 'pre-terrorist' days the security police were constantly warning about sabotage. Of course they were correct in creating heightened awareness of a potential threat.

133

However they went to extremes. They had little red blocks of wood with the word "bomb" on it. They would hide these in crew vehicles and if you didn't find it, it was a ding against the crew. So anytime you left the facility and parked you had to check for a little red wood brick before you moved the truck.

That got to be a pain in the butt at times. One morning we all went off to the trainers for about an hour session. We finished about the time the alert kitchen opened and really wanted to get back for lunch. None-the-less, we dutifully inspected the truck and found our bomb laying on the engine block. So we followed procedures and alerted the Security Police. They responded and we did the paper reports, etc. We did it right but lost time getting to lunch. Of course the whole crew bitched about it on the ride back to the alert facility.

When we got there the radar saw a red butt can on the ramp that led into the building. He picked it up and wrote "bomb" on it with a grease pencil. About a half-hour later the security cops arrived and surrounded the area. In the aftermath, all the crews were assembled in the briefing room. The Ops Officer and Alert Force Den Mother were there with the head of the security cops - AND the red butt-can bomb. Rightfully, they were not amused. The usual, "Confess to who did this and there will be no punishment" pitch was made. No one volunteered any information. So the mystery remained. But the result was positive. Prior to the butt-can bomb the SP hidden bombs were almost a daily event. They tapered off to about once a tour after that. This little prank wasn't wise by any means. There was a possibility of sabotage and the Security Police were just doing their job. But Crewdogs don't always show great wisdom when it comes to pranks. Dave Lowell was the culprit, so if anyone recalls the incident – you now know who "dunnit."

A Boeing assembly line in the alert shack was entertainment for one tour. The Monogram B-52D 1/72nd scale model kit had just hit the shops. All the crew wanted one so I agreed to supervise the building. If I remember right we bought five of them and after all the usual daily activities we would go to one of the planning rooms and work on them. I have been a model-builder most of my life, so I became the instructor. Each guy took some stage of assembly and away we went. I did most of the pre-painting of parts with an airbrush and everyone else glued nacelles, wings, fuselage, etc. It all went very well and in four days of a seven day tour we were ready to roll them out. The fun part was taking them out to the ramp and establishing a perspective shot to show

D-models on the Dyess Ramp. As an aside, I also built a B-52 model while flying Chrome Dome. It was the 1/144th scale Revell kit. I obviously couldn't paint it, but with odorless glue I did assemble it all in-flight and finished the painting at home.

At Dyess, we were also a 'flying bone-yard" and all the E-models wound up there. For the most part they just sat there, but I think they had to make one flight every 90 days or at some other interval which I have now forgotten. Most crews would allow much longer preflight on those airplanes. The ground crews did an outstanding job of cleaning out the bird nests, checking fluids, and otherwise getting the old E ready to fly. On one of those scheduled flight our crew found a particular BUFF had a lot of problems. First there were electrical problems. We resolved those and then found something else. I think it was some of the hydraulic packs were DNIF. Maintenance came out and worked without success. Finally the AC decided it was a no-go. One of the Wing Staff Colonels showed up and told the AC if he wouldn't fly it, he, the Colonel, would take the mission. The rest of us had started downloading our gear and the Colonel said to load it up again. We all said if the AC won't take it – we won't either. We all feared our flying days were over, but the Colonel didn't press the issue and that flight didn't launch.

The final conflict concerned the ORI. Of course there was a betting pool and it wound up being a pretty good sum of money. To make a long story short, our crew had the best bombs, navigation, and ECM run and won the pool. We were elated and then deflated. We were rather rudely presented the money with no congratulations, no thanks for a good job, just a Colonel walking over and handing the money to the AC. We didn't care about the money at all. In fact, we hosted a post-ORI party at the club - for crew members only!

In spite of these tales of woe, Dyess was a good experience. The Crewdogs there all hung together and we enjoyed great camaraderie with them. Dyess marked the end of our crew (I think it was E-13). People shipped out to other bases or schools and we never came together again. So it was another base, another BUFF, and on to other opportunities.

Was It All for Naught?
Derek H. Detjen

During the height of the Cold War, the Russian threat and the ever-present possibility of the onset of WW III were very real. The unpredictability of Nikita Khrushchev, the Cuban Missile Crisis, and the heightened international tensions all made our endless alert tours with the B-52 far from boring. Our crew was actually airborne and on the way east during the time that the first Russian ships approached the naval blockade around Cuba. We halfway expected to get the "Go Code" during one of those missions. Every alert facility had teletype hook-ups to both the AP and UPI news services; ours was in the library room and was monitored regularly by all the crewmembers. That library would figure later on in an incident that I've remembered for a lifetime.

There was also a day on alert that we were sure that we were going to realize our worst fears and launch on our nuclear deterrent mission. It was in the late afternoon and I think there was a fast-pitch softball game in progress on the alert diamond when the klaxon went off for the first time. I remember running full speed up the Christmas tree in my baseball spikes to reach our aircraft, fortunately parked in one of the first two parking stubs. The rest of the crew arrived in the alert truck and we boarded the plane to find that it was just a locally directed, non-taxiing "Bravo" exercise. It was dinnertime by the time we re-cocked and buttoned up the aircraft and went back inside.

Later that evening, we were engaged in a serious bridge game in the Officer's Lounge as midnight approached. Capt Ed Noyallis, the EW on another crew, had just volunteered to go to the dining room, where we had midnight chow every night, and returned with a tray full of coffee cups, slices of cake, and ice cream. Just as he began to set the tray down, the klaxon went off again, someone bumped Ed from behind and the coffee, cake, and ice cream flew everywhere, including all over a lot of flight suits! Everyone vacated the building, hit the trucks and before we knew it, we had started engines and taxied out to the runway on a "Coco" exercise. By the time we had recovered, returning all the aircraft to their original parking stubs and topped off all the fuel tanks, it was in the wee small hours. Sometime just before 4 a.m. the klaxon went off again, and we just knew that WW III had just begun! Fortunately for everyone, it was once again a "Bravo" exercise, this time initiated by our 42nd Air Division at Seymour-Johnson. Needless to say, we were all relieved when we finally hit the sack that night!

It was on another evening at the Turner alert pad that several Crewdogs were sitting around in the library, just telling war stories and shooting the breeze. 1st Lt. Jimmy Condon, then a navigator on another crew, opined that "if we ever go down the tubes, it won't be the Russians or Red Chinese who'll do us in but rather, we'll do it to ourselves, from within, without a shot being fired!" I never forgot Jimmy's dire prediction, remembering it for the rest of my Air Force career and subsequently. Last September, Betty and I went to Branson, Missouri, for an absolutely incredible Arc Light-Young Tiger Reunion. The hotel was jammed with old Crewdogs, almost all of us still married to our original wives, moot testimony to SAC, our families, and the military camaraderie that we all shared. Generals LeMay and Powers would have been proud of that gathering, for sure.

I was happily surprised to meet Lt Col (Ret.) Jimmy and Jennie Condon there. He had been shot down over Hanoi during a Linebacker II mission, and fortunately spent only about three months in the Hanoi Hilton before he was repatriated. Jimmy and his wife conducted that special little ceremony honoring our missing airmen preceding our dinner that first evening at Branson. I reminded him of his prediction regarding the ultimate fate of our country, made years ago. We agreed that we'd lived long enough to see it happening before our very eyes. How terribly sad, but every time I mention writing my book about my years in SAC titled "All for Naught," Betty assures me that "everything all of you guys did bought us many more years of freedom," which is how we must look at it in the overall scheme of things.

I'm sure of one thing. I haven't had the feelings of camaraderie for many years that we experienced at Branson last year. The SAC crew force was by far the most highly motivated, professional group of folks I've ever had the pleasure of calling friends, and I don't think that any civilian could ever experience the kind of mission commitment that all of us had. I still feel proud to have been a part of it.

Chapter Three

outheast Asia [south-eest] [ey-zhuh] - *noun* - The countries and land area of Brunei, Burma, Cambodia, Indonesia, Laos, Malaysia, the Philippines, Singapore, Thailand, and Vietnam.

We Were Almost Heroes
David R. Volker

How can a Crewdog become a hero? Sometimes all you have to do is what you are assigned to do, and do it under extreme circumstances. Sometimes you have to die performing your assignment to qualify for this possibly enviable title. There were many heroes among the Crewdogs who went to war in Southeast Asia. They achieved hero status honorably while doing their jobs professionally. They did the best they could, considering what they had to work with: old aircraft, WWII weapons and tactics, and a temporary duty (TDY) mentality.

In this particular case, despite an incredible string of errors by all concerned that should have led to our deaths (and possible classification as heroes), we survived the first combat encounter with a MiG-21. Certainly, if Sgt. Sid L. had been able to shoot down that MiG, he would have been a hero....

A close encounter of the fourth kind... (shots fired)

The three B-52 D-model aircraft that made up Copper flight took off from U-Tapao Royal Thai Navy Airfield just after midnight local time and climbed heavily laden with weapons and fuel above the overcast without much incident. My pilot, Capt. J., was flight leader since he was the most senior aircraft commander. He remained flight leader throughout the rest of the flight. He was to lead the 18 men and three aircraft on an historic and almost ill-fated mission that early morning, but we were all unaware of what was about to transpire.

Our aircraft, Copper 1, began experiencing mechanical problems soon after our climb out was over. First, our gunner, Sgt. Sid L., informed us that his gunnery radar was seriously degraded. He was unable to reliably track the other aircraft in the cell. Then I found that the #2 UHF radio did not receive properly. We were then down to one UHF radio for inter-plane communications and air-to-ground communications. Because of these difficulties, my pilot decided to inform the other aircraft in the cell to prepare to perform a lead-change maneuver. There was little time to do so as we were rapidly approaching enemy territory and our target for that evening, the Ho Chi Minh Trail at Ban Karai Pass.

We accomplished the lead change successfully and took up the number two position in the flight while retaining command authority.

The copilot in the lead aircraft then had responsibility for most of the communications that I once had. I knew that he was inexperienced to the point of being unreliable and bore close watching. Perhaps I should have watched my aircraft commander more closely as well. As we approached the line defining enemy territory, I ran the appropriate checklist. One of the items on that checklist called for the pilot to turn off all exterior lights. I did not receive a verbal confirmation or response on this item from him over the intercom. I shouted across the cockpit to get his attention and to remind him of the lights and his affirmative nod was all I saw. We were not allowed to review that portion of the intercom tape after the flight. Little did I know that he allowed the rotating beacons to remain on for the duration of the flight. During debriefing later the crew of Copper 3 confirmed this fact. As a result, we were the only aircraft anywhere over Laos or Vietnam that night with lights on.

The flight proceeded without incident to the target area. The air above 35,000 feet was clear and smooth. There was a broken undercast below us and I remember looking up and admiring the stars of Orion's belt through the window above my seat. There was little chatter on our radios because the noisiest frequency, GCI, was usually received by our #2 UHF and that radio was out. The #1 UHF was set to Bombplot frequency and we were headed east, IP inbound. Bombplot was a highly accurate radar site on the ground that would track us carefully, give us minor course corrections and eventually tell us when to release our bombs. We in Copper 2 would not be able to hear threat warnings directly but would instead rely on our lead aircraft to relay them to us on Bombplot frequency, which also served as inter-plane communication frequency.

We were in good Drift Angle Station Keeping (DASK) position behind the lead aircraft, Copper 1. Lit by starlight from above, I could easily see the outline of his aircraft against the clouds below. The dim red glow of his engine exhausts was also visible against the blackness

141

of his outline. We were all about to open our bomb bay doors and perform our High Altitude Plowing System (HAPS) mission for the evening over the Ho Chi Min Trail. Bombplot was giving us small corrections to line us up with the target. Suddenly over the UHF came the voice of the lead pilot or copilot saying "Divert, divert!" We were less than 30 seconds from the target and that was quite unusual. He called his turn and immediately entered a 45-degree right bank. According to then current doctrine, he would roll out after 90 degrees of turn was completed. After counting about 15 seconds, my pilot entered the same maneuver. We were still fully loaded with 108, 500-pound general-purpose bombs in our belly and under our wings.

Unknown to us to that point, a North Vietnamese MiG-21 had been launched a few minutes before from a nearby base. Later we learned that his flight path followed a carefully practiced route. During the past few nights, when North Vietnamese radars detected an incoming flight of B-52s, the enemy pilot timed his take off and climb out to show himself on our GCI radar at just the right moment. Our GCI controllers then transmitted the MiG threat code to the B-52 cell and, usually after just a few seconds, the cell turned south. This maneuver on his part saved his troops on the ground and also served to perfect his timing. His eventual goal was to bag one or more B-52s without having to work too hard or expose himself to possible interception. In fact, if his timing was just right and the B-52s performed the standard divert maneuver in a predictable manner, the MiG pilot arriving from the north and heading south would position him immediately behind and below the cell. This is perfect firing position for his Atoll heat-seeking missile.

The early morning of November 20, 1971, turned out to be a good news/bad news night for the MiG driver. The good news was that he didn't need to use his radar to find our cell. Two bright red, flashing rotating beacons on our belly were easily visible against a moonless sea of stars. All he had to do was look up and line up the nose of his plane and climb. The bad news was that we did not follow a predictable course for him. Every other cell that he had ever approached had turned away from the target within a set period of time after he was an item of interest on our GCI radar.

This time it was different, the copilot of Copper 1 was inexperienced. He had never had the responsibility of managing the communications for an entire flight. It took him over a minute to recognize the divert word for our flight that night. "Snowblind" repeated three times over the GCI frequency by the controllers just did not sink in very fast. After it finally registered, he confirmed it with his pilot, further delaying the divert maneuver. This delay had the effect of throwing off the timing of the MiG intercept. But the MiG pilot was undeterred; after all, he had a visual target! He would not need vectors from his GCI. He never asked for them and they were never offered. He would also not need to use his air intercept radar to close with a visual target. None of the electronic warfare officers in our flight ever detected an AI signal from a MiG. Nevertheless, the MiG-21 pilot was forced to perform a series of violent S-turns to line up behind Copper flight, while he climbed to our altitude on afterburner.

All he could see was the beckoning red flashes of Copper 2. He was locked on; ready to fire his Atoll heat seeking air-to-air missile, when looming out of the blackness was a patch of even darker black, silhouetted against the stars. He almost collided with Copper 3, the last aircraft in the cell. Whether he fired his missile during his avoidance maneuver or immediately thereafter, I don't know. Both the tail gunner and the copilot of Copper 3 saw him go by as a blue streak, afterburner glowing brightly. Neither of them bothered to tell anyone else in the flight what they saw until after they landed. The NVAF Atoll was on its way, trying to home on the heat signature of our engine exhaust.

Meanwhile, we in Copper 2 had nearly completed our turn and were about to roll out of our 45-degree bank and take up a southerly heading. Just before we rolled wings level, a loud explosion and bright flash of light came from the left side of the airplane. Capt. J. grabbed for the control column and began to struggle with controlling the aircraft. He had flown the turn on autopilot and the explosion turbulence had knocked the autopilot off. I checked the engines and found they were all working just fine. I also noted that we were holding cabin pressure and that the airplane seemed to be pretty much in one piece. Then I heard my pilot say those fateful words "Mayday, Mayday!" over the UHF. He told me later that he had been in the first B-52 to be shot at by a SAM on his previous tour of duty and thought it was happening again. He was still struggling with the aircraft, perhaps out of panic or just poor pilotage. He couldn't seem to hold his altitude or heading properly. I'm sure he thought he was having control problems caused by battle damage. He asked for full power and I slammed all eight throttles the remaining half-inch to the firewall. Just after that, he told the crew over the intercom "Prepare for possible bailout!" I really didn't know exactly why he uttered those words at that time, but I can assure you it really got the crew's attention. During this time of confusion and fear, I took a few seconds to look out my windows. At two o'clock level I saw a beautiful blue flame receding into the distance in front of the aircraft. As I watched, it went down and then to the rear under our right wing. I keyed the UHF radio and reported "Bandits, two o'clock low!" It was not until several days later that I learned that flame was probably the afterburner of the MiG-21 that had just shot at us and was now "split-essing" for home at low altitude. It is interesting to note that during this entire engagement, I was the only crewmember in the entire cell to report a sighting over the radio so that others in our cell could hear it.

By that time things had quieted down in the cockpit and the pilot seemed to be regaining control of the aircraft. We found ourselves straight and level at 36,500 feet, on a southerly course. The other members of the flight checked in on UHF and all were OK, but not necessarily in proper formation position. Suddenly the gunner from Copper 3 or our gunner reported aircraft approaching from our stern. Capt. J., our pilot, was still in command of the flight even though he was flying the number two plane.

He ordered Copper 1 to begin an evasive right turn (45 degrees of bank, 90 degrees of turn) on his command. Since this was the only authorized evasive maneuver we were allowed and since the "threat aircraft" were approaching from the right, Copper 1 turned when he was told. We turned 15 seconds later and Copper 3 turned 15 seconds after us. We all rolled out, in trail, headed west, back toward Thailand. A few seconds after rollout our gunner reported more aircraft approaching our formation. I knew if the pilot ordered another turn, we would end up headed back toward North Vietnam. Instead the pilot proposed an incredibly ill-conceived plan. He told the other members of the flight that it appeared that we, Copper 2, were the target of the attack and that if the approaching aircraft got closer, we would break out of formation and lead the attackers away!

It occurred to me at this point that since I could not hear GCI, and did not know what was going on in that important area, our "attackers" may in fact be friendly MIGCAP aircraft. I quickly switched our UHF radio to guard channel and said "Aircraft approaching Copper 2 and Copper flight, you are coming within gun range. Break off now or you will be shot down!" in the most authoritative voice I could muster given my fear.

Within seconds after my transmission, the aircraft that had been approaching broke off and dropped from our radar screens. We all took a deep breath and paused to assess our damage, if any. None of us in the flight reported damage or hits. We regained our formation and proper spacing and went on to bomb our secondary target about a half-hour later.

I was happy to get rid of our bombs at last. Soon after bomb release we prepared for our descent and landing back at U-Tapao. As I ran the pre-descent checklist and examined the various aircraft systems that I was responsible for, I took my fuel readings. I immediately noticed that the left drop tank was almost empty. We never used fuel from these tanks because having fuel in them helped improve the stability of the aircraft and prevented wing flexing at high speeds. I consulted with Captain J. about the apparent loss of fuel from this tank and we decided to balance the aircraft by transferring fuel out of the right drop tank into the various main wing tanks. The actual loss of fuel was not serious, only about 9,500 pounds out of our total fuel load. I performed the transfer as we descended and by the time we reached approach altitude we were balanced. We did not declare an emergency, though in retrospect, we should have.

We parked the airplane after an uneventful landing. After we exited down the hatch, we gathered near the left wing tip to see if there was any visible damage. The wings didn't droop very much as they had been relieved of most of their load. We found the tail cone of the left drop tank bent downward at about a 30 degree angle. This was just enough to crease the skin of the tank and allow the fuel to leak out. Wing staff subsequently assured us that what we saw was not battle damage, but rather the result of a malfunctioning pressure relief valve in the tank. Maintenance assured us that this was physically impossible given the design of the drop tank pressure relief system, but Wing staff is always right. To have admitted that this was battle damage would have raised many questions with only embarrassing answers.

I remember very little about our debriefing as it was so uninformative. We asked more questions than the intelligence staff did. We answered theirs to the best of our ability with complete descriptions of what we saw and did. They answered ours with shrugs and excuses.

To characterize the events that lead up to our near deaths as an unfortunate string of screw-ups is an understatement. The phrase "Shot at and missed, shit on and hit" aptly describes how our crew and the others that flew with us that day felt. We were considered anything but heroes.

What should have been a situation all crews and staff could have learned from instead was swept under the rug with only passing reference given to it in future training.

Perhaps the whole incident was just too embarrassing:

To the staff at U-Tapao who failed to warn us of the threat pattern that had been set up by the MiG,

To my pilot who failed to turn off the rotating beacons of our aircraft making us a target in the sky,

To the pilots of Copper 3 who saw our rotating beacons were on in enemy territory and failed to warn us,

To the copilot and pilot of Copper 1 who didn't realize that they had received the divert word until it was almost too late,

To the gunner and copilot of Copper 3 who saw the MiG but didn't report it to the flight,

To the copilot of Copper 3 who was so rattled by the whole experience that he requested and got a one week R&R back to the States,

To my pilot who panicked and nearly commanded his crew to bailout of an effectively undamaged airplane,

To my pilot who nearly broke from the protection of the formation when he perceived a threat,

To GCI who failed to inform the flight that MIGCAP aircraft were coming to our aid and almost got them shot down for their trouble,

To the entire theater Air Staff who failed to learn from all of our mistakes and sent dozens of B-52s into Hanoi without proper planning, protection, or procedures a few short months later.

Footnote:

One indication of the lack of learning is the quotation from Major Billy Lyons, aircraft commander and cell leader of Lilac 1 when he says in his report of his trip over Hanoi "...The sight as we turned over the IP inbound to our target is one I'll never forget. The red rotating beacons of the three-ship cell two minutes ahead of us stood out vividly against the dark night." (Emphasis added) This quote is taken from page 95 of reference #1 and, I think, illustrates my point very well.

Ref. #1 Lt. Col. Karl J. Eschmann: "LINEBACKER-The untold story of the air raids over North Vietnam" 1989

Ref. #2 Beaton, Col. Clifford M. End of Tour Report: 7th AF Intelligence, April 1971-April 1973, Corona Harvest Program, 20 July 1972. SECRET, declassified 31 Dec. 1980, p. 19.

Linebacker II - Straw 2 Bravo
Jim Farmer

In April of 1972 our March AFB B-52D Crew, designated E-10, was told that our 26-hour training flight orbiting Alaska scheduled in two days had been canceled. We were to be prepared to deploy to anywhere in the world with 30 minutes. Two days later our squadron was flying all its BUFFs to Guam. A couple of days after that we were flying combat missions to South Vietnam. That was our introduction to "Bullet Shot." Our Crew was comprised of Capt Deverl Johnson, A/C; 1st Lt Jim Farmer, Copilot; Maj Frank Gould, Radar Navigator; Capt Vince Russo, Navigator; Capt Paul Fairbanks, EW; and TSgt Walt "Budda" Barcliff, tail gunner.

Deverl Johnson, who had earned his commission as pre-academy Air Force cadet, was a former F-4 navigator, and had spent a year in Vietnam flying 0-2's before assuming his B-52 responsibilities. He was a devout Mormon with four children and was dedicated to his family. Deverl was a good pilot and an excellent Aircraft Commander.

Frank Gould, at 39, was the oldest member of the crew. Frank was a New Yorker, sharp as tack, with a wonderful sense of humor. He was proud that his radar photos of the first night were presented as examples of how to do it right for the third night's briefing.

Vince Russo is a consummate professional. He was completing his second master's degree during his Bullet Shot tour. He continued his flying in the reserves while becoming a civilian military analyst for the Government in D.C.

Paul Fairbanks was a high school math teacher for three years before joining the Air Force. He did not have much luck in interesting the Air Force Academy in his desire to teach there prior to Linebacker II. Afterwards, the Air Force sent him to earn his masters and PHD in math. He spent the remainder of his career teaching at the Academy.

Walt Barclif was a gunner's gunner - always ready to do whatever need to be done. He loved Thailand.

As a crew we had been together over a year before Bullet Shot, pulling alert, training, and flying. We were a SOLID crew professionally and personally.

Linebacker II came towards the end of our second 120-day tour. We had flow both out of Guam and U-Tapao. As we had been flying "press-on" missions to the North the previous month from Thailand, we had experienced what it like to fly in the vicinity of active SAMs. We were not happy about those missions; the targets' significance never seemed high enough for us to be hanging our asses out for them. I didn't want to die over a suspected truck park.

In December we rotated back to Guam. Paul and I had had our wives with us in Thailand. They/we stayed at a secure facility on Patta Beach. They joined us on Guam as well. The benefit to being on Guam was that we no longer had to fly missions to the North since they were done out of U-Tapao

We knew something very big was up prior to the first Linebacker II briefing. Rotations were canceled. The base was shut down and missions stood down. That had never happened before. There was to be a massive briefing, scores of crews, not just the usual three. It didn't take much to have an idea of what was about to happen. We were not really shocked when the commander announced that our target was Hanoi. There were two emotions experienced by each the crewmembers assembled in the room at that time. First was in response to the obvious danger of flying straight and level into the most heavily defended city in the world. The second was the pride and motivation associated with doing something that might very well result in our fellow aviators being released from their years of captivity in the Hanoi Hilton. By God, this was a mission worth hanging your ass out for, finally!

I don't remember much of the first evenings mission, other than there was a lot of radio chatter, more SAMs then ever before, and lots of parachute beepers going off. Deveral saw a BUFF off our 9:00 go down looking like someone pouring a pitcher of fire out of the night's sky. We and our cell made it in and out of the target on our 14 ½ hour mission in one piece. The line of contrails from all the BUFFs in front of us on the way home was a sight to be seen.

We were in crew rest for the second night's missions.

The third night's briefing for me was more onerous. I remember my good friend Mike Kenny (a tall, Texan copilot who is not unlike Slim Pickens in the Movie "*Dr. Strangelove*") wearing his cobra skin cowboy boots to the briefing. We were to be Straw 2. The second aircraft in the second of three cells, hitting exactly the same target we had hit the first night, using exactly the same flight plan. We had been briefed that the SAM sites liked to figure out what was going on by finding and observing number one and then pick off number two. Second aircraft, in the second cell - not good! Additionally during the first night's briefing we were told that this was to be a three day push. I assumed that if I knew that, so did the enemy and he would pull out all the stops that night.

Anticipating that our chances of getting hit were strong, I put a pair of socks, extra water, a ham sandwich, candy bars and the survival knife I had purchased in survival school in my flight suit. If I did survive getting shot down, those would be useful while evading the enemy.

Watching and listening to a base full of B-52s taking off at one minute intervals for two hours is another sight that is not to be missed. Everyone was a pro! From the guy that packed the lunches, to the guys that loaded the bombs, to the guys that turned the wrenches to Charlie tower who orchestrated the launch. It was something!

The flight to the target was long and tedious. I remember being very tired ½ hour out from the target. Adrenalin, and years of exacting training would take care of that. Sure enough! Going into the target that night was going to be hot. SAMs were going off everywhere. The night was clear, except for a low cloud deck. The SAMs lighting off were very interesting, almost beautiful to see. At first they were a broad dull light until they pierced through the clouds. Then they changed to a large bright light flying up from down below. Almost all of them seemed to be shot ballisticly. That is, they didn't appear to be guided or changing directions. All except for the one that got us - that one definitely was being guided.

We were straight and level over the target and just about to start bomb release when it lit off. It was on its way when we initiated bomb release. The EW called uplink indicating we were the target and it was being guided. We could not evade - you did not want to take the chance of tossing a bomb into the Hilton. It was apparent to me that we were going to get hit. I announced it to the crew. The missile exploded about one second after our last bomb released. I'm sure they passed each other in the air. We initiated our post target turn and I believe the gunner had significant wounds when he reported more SAMs were being sent our way. The pilot, navigator, and radar navigator also had sustained significant wounds. I had a small one, but didn't realize it. The good news was that our wings were still on and we were not on fire.

Immediately after the post target turn the pilot assumed a best range airspeed (predetermined at 215 kts indicated) and the flight plan egress heading by using the whiskey compass illuminated by a flashlight. The engines that were working were being fed by gravity. At that point we had about a 500/mm rate of descent to maintain the 215 kts. I tried to get out a "mayday" on guard by opening my window (yes, unbelievable isn't it) at 40,000 ft and transmitting on my PRC-90 survival radio. I later discovered that I was talking into the speaker but it was dark and I couldn't see the difference. At that point we were very busy keeping the plane flying. We were able to do that because we were flying a D-model. It had inboard ailerons in addition to the main control surface, the spoilers (which were mop from battle damage). The D-model had these normally insignificant flight controls (about the size of a garage door) which were controlled by a trim tab about the size of a shoe box top. They were connected by wire cables to the pilot's controls. So there we were, flying this monster with a control surface about the size of a large shoe. It worked though.

What really was amazing was how calm and professional everyone was during the time before bailout, which probably lasted 20 minutes. The people that had things to do, did them. Those that didn't, kept quiet. Frank who was the most injured complained about the inadequacy of the first aid kit. He said that he was going to write it up when we got back. It was a comment right out of *Catch-22*. I miss that guy. I remember thinking that I won't have to pay for a stop watch that I been issued but had lost. The ability to deal with such a set of circumstances effectively is a testament to all the training we each received.

We were streaming fuel from the holes inflicted by the SAM. As engines quit we had to increase our rate of decent. We were tiring to make it back to friendly territory in Thailand; however, at 20,000 feet, with a 1,500 ft / mm rate of decent and having difficulty maintaining wings level, the Aircraft Commander gave the order to bail out. We knew that we had not crossed the Mekong River. We were going out in enemy territory.

We had a controlled bailout situation. We went out in an orderly and sequential fashion. Earlier that month an EW and copilot ejected simultaneously and their seats collided and someone died. The tail gunner goes first. He pulls a lever causing the entire radar and machine gun assembly to fall off; they he just steps out. The Navigator is next, ejecting down. Then the EW, ejecting up, followed by the Radar ejecting down. Finally the copilot and pilot eject up. Given the time and airspeed involved the crew was spread over four to five miles in a straight line.

The pilot and I heard the Navigator's, EW's, and Radar's seats fire. I then turned to Deverl and said "See ya' round", straightened my back and squeezed the trigger. We went out between 16,000 and 17,000 feet. Deverl later talked about how weird it was to be flying a B-52 solo with a big empty hole where the copilot used to be.

The seat worked properly and I waited for my chute to open automatically at or below 14,000 feet. Consequently the next thing I knew was that I was free falling. It was pitch black out. I was holding on to my helmet with both hands as we had been briefed that guys had problems loosing them during the ejection process. It was a good thing to have on when landing through the jungle canopy. Just as I was attempting to stabilize my fall by getting into the spread eagle position my chute deployed with a real hard thump. Looking up I could see that two of my 28 panels were shredded. Not to worry. It was cold at that altitude.

While hanging in the chute a HUGE fireball exploded below me. The BUFF had impacted the ground. When its pitch dark outside and there is only one light source it is impossible to judge distance. I was convinced that I would land in the fireball. That was scary. I pulled on the opposite riser attempting to head in the other direction. Actually the plane was probably miles away.

I have no idea how long it took to reach the ground. It felt like a long time. By the time I approached the ground I was getting first light. I could make out a ground cloud deck or fog below me. I prepared for a tree penetration - keep your legs together and reduce exposure to your chin by turning your head.

I could feel and hear myself passing through the jungle canopy. Then I was on the ground with my chute still in the trees. I was on a slope, dense with trees - almost forest like. It was dry which surprised me since I had always expected the jungle to be wet. I was much relieved that I wasn't injured. I immediately drank some water to ward off shock and examined my survival kit to see what might be useful. Just then I heard a low flying jet. I proceeded to send out a "mayday" on my survival radio. It was Paula 4, an F-4, probably on his way back form escorting the BUFFs. He asked me my call sign and asked if OK. I indicated that I was Code 1, which is what we'd call to maintenance if everything was working fine. He advised me that he'd relay my message and position, then suggested that I head for the high ground. That seemed like an excellent suggestion as I thought I heard activity downhill - perhaps vehicles   I ran for a long time uphill. Finally I found a hiding place among the vegetation and laid down, hidden by a log and waited for the rescue forces. I was sure they would be coming soon.

It wasn't too long, perhaps an hour, before there was a Sandy (an A-7 search and rescue fighter) calling for me on the survival radio. I had inserted the radio's earpiece, which enable me to listen without making a lot of noise. I assume I was also whispering into the mike as I was concerned about attracting the attention of anyone that might be around. The Sandy made two passes 90 degrees from each other. Per his instructions I gave him a hack as the Doppler effect indicated that he was passing by. He indicated that he thought he knew which ridge I was on, would relay my position to the chopper and for me to maintain my position if I could. Then he was gone.

Perhaps an hour or two later I heard the Jolly Green Giant (a huge Air Force rescue helicopter) come up on the rescue frequency. They were coming to get me. I was still hidden under the log. I didn't dare look around. Using my survival compass and the sound of the helicopter I proceeded to vector the chopper towards my location. The chopper sounded pretty close when he indicated that there was a problem. The ground fog prevented him from seeing the ground. That meant that he couldn't descend or lower the rescue device. He made several attempts to get me but not being able to see the ground prevented him from doing so. Then he said that there was a hole directly below him. If I could make my way to him he should be able to get me. He sounded about 100 yards away. So I went for it, running as fast as I could, not looking around to see if I was being observed, making a hell of a lot of noise as I crashed my way through the jungle. Running toward the sound of the aircraft I finally made my way to where it was directly overhead, but I couldn't see it through ground fog. As I was running the hole had closed up. He couldn't do anything for me. In addition to that he indicated that he was low on fuel and had to leave to gas up. If there were any bad guys around I had just announced my position and the helicopter which was directly above me was about to go away   THAT moment was the emotional low point of my life.

At that point Echo (the Electronics Warfare Officer) came up on the rescue frequency, and told the helicopter that he's on a ridge nearby above the ground cover. Would they swing by and get him on their way out? That's what they proceeded to do.   Now, Paul and I are good friends. We did a lot together. Our wives were back at the base together sharing an apartment. We are both competitive people but it seemed whenever we competed against each other, he would somehow edge me out. It didn't matter whether it was tennis, basketball, chess, ping pong, whatever; he'd find a way to edge me out. We once had a tennis game go 14-16, he won.

At that point I'm scared as hell and pissed at Paul because that SOB may just edge me out again   Fortunately there didn't seem to be any of the enemy around   About an hour later the chopper returned and hoisted me up with out incident   When I got pulled into the chopper Paul was sitting there with a big you-know-what kind of grin on his face. A second chopper had picked up the Pilot, Navigator and was searching for the Radar. Ours started to look for the gunner. A PJ had to go on the ground to find him and get him hoisted up   We were told that the other chopper was in contact with the Radar. He was complaining that he was missing his Martini hour (which is just what Frank would have said) so we headed back to the chopper's base.

Our arrival at NKP was a big deal - lots of people, lots of brass, someone opened a bottle of champagne and I could not hear a damn thing after three hours in the Jolly Green Giant. SAC had a KC-135 right there to wisk us back to their control. We didn't get a chance to appropriately thank our rescuers, nor did we have a chance to talk to the people coordinating the search for Frank. Perhaps we could have told them that we went out in a straight line and they might look between where the EW and copilot were found.

The other chopper took the Pilot and Nav to another base where they were hospitalized. Paul, Walt, and I were flown to U-Tapao. Walt stayed there while Paul and I went back to Anderson AFB on Guam.

As I mentioned earlier, our wives were on Guam at the time. Fortunately by the time the Air Force had found them, we had already been rescued and were on our way back. Paul and I were asked to brief the brass. We were NOT to come in flight suits. I had to borrow a blue uniform from my buddy Mike. "Welcome back to SAC's idea of combat reality." The briefing room was large. I had never seen so many full colonels in my life. During the course of the briefing I described how dangerous and unexpected the downwash of the helicopter was with tree limbs falling all around me. It would have been good to have my helmet on. When asked why I didn't, my response was that SAC required them to be a shinny white. I didn't think running around the jungle like that was a good idea. Some full colonel asked me why I didn't apply colored bug repellent to it. I don't believe that I answered him. The next day, all helmets in SAC (including stateside) were camouflaged.

With the Pilot and Nay in a hospital in Thailand, the Radar missing and the gunner probably still in Thailand, Paul and I helped pack Frank's personal items. SAC sent us back to March AFB, California. We arrived on December 24th. That evening I went to Frank Gould's home to explain to his wife and 14-year-old daughter what I knew. The next day, Christmas, I flew on a 747 from LA to my family's home in New York. The flight was so empty it had as many flight attendants as passengers. I felt very guilty that holiday as I was home and my friends were still flying missions, and getting shot at.

There was nothing to do back at March since all the planes were overseas. However, the prisoners were soon released and March was one the bases to which they returned. It was wonderful to follow their release on TV. I felt honored to be present at the time they stepped of their plane, back on U.S. soil (actually tarmac). My buddy Mike rotated back and proceeded to break both legs while skiing. Mike went on to be the first operational B-1 pilot, a below-the-zone full Colonel, and Director of Operations at Ellsworth with flight responsibilities for two B-1 and two KC-10 squadrons. His crew, which was the B-52 Tactical Evaluation crew for Southeast Asia needed a copilot, so I went with them to Thailand. From the media's point of view the air war in SEA was over when the prisoners came home; however, we were still flying combat missions daily out of U-Tapao. That's where I understood the meaning of Frank's expression "a complaining soldier is a happy soldier." With everyone else gone or going home, we were still there on open-ended continuous "temporary duty". Morale was so low no one bothered to bring it up in conversation; however, one day a C-141 with a hydraulic problem diverted to U-Tapao for overnight repairs. Its passengers were the A-7 Sandy squadron rotating back to the states. Needless to say that was a very exciting and expensive night for me in the O' Club. It was worth every penny. Not only did I meet the entire squadron, I talked at length with Arnie Clarke, Sandy One, lead of the rescue mission. He's the guy I whispered to from under a log just a couple of months earlier. I've seen Arnie several times recently. We each have a recreation place near each other in the Cascade Mountains in Washington State. He's a true hero with several tours, shot down twice himself and a recipient of the Air Force Cross.

There is another interesting aside to this story. A couple of years later at my 10th high school reunion I was interested in what had happened to a girl I had known. In talking with her brother I learned that she had married a guy who was an Air Force A-7 pilot. He was in the squadron responsible for our rescue. I thought, by golly, I'll bet I bought him a drink or more in U-Tapao without ever knowing the connection. Many years later when they were guests at our home, we discovered that he was in charge of the A-7 Command Post that night. When he got a call to launch the A-7's, Capt Steve Donahue got into a heated discussion with some general by telling him that they couldn't launch until first light. Sandies couldn't do SAR at night. What a small world!

150 Foot Bag Drag Times Four
or
Getting a Bonus Deal the Hard Way
George Schryer

On 21 October, 1969, crew E-21 from Seymour Johnson was just settling into the routine of flying combat missions over South Vietnam. We had been on Guam for almost two months and were getting comfortable flying in the D-models after retraining from the G-models we had at home. The crew had been together for about a year so we had learned to work together quite well.

On this particular day we were tagged as number two in the cell and as the mission proceeded through preflight, taxi, and takeoff everything was routine with the exception that on this mission we had a Staff Flight Check Officer aboard. At about 10,000 feet I noticed that my cabin pressurization and heating system was not working. I notified the A/C Capt. Duelfer of my situation. I told him I had my cold weather flight suit available if he wanted me to stay or I could come forward as was proscribed in the manuals. He and the Staff Officer decided I should come forward before we got to the refueling area but as the SO had the only extra parachute I would have to make the move with my own.

After setting up the compartment for my absence and notifying the navigator, I started my trek. The Nav was to monitor my progress through the bomb bay by looking through a window in the bulkhead that separated the forward crew compartment from the forward wheel well and bomb bay. I grabbed my walk-around oxygen bottle and crawled into the 47 section, which was the area directly forward of the gunners compartment and was where all the aircraft hydraulic machinery was located. So far so good.

Now for the bad part of the journey. The walk-way along the starboard side of the bomb bay was not designed for a leisurely stroll. It is only about 18 inches wide and for some arcane reason it is covered with aircraft rivets. Unfortunately the heads of the rivets are on the underside of the walkway and the collars of the rivets are on the top side of the walkway. That is not a problem if you could walk upright but the only way you can make this trip is on your hands and knees.

Now the gunners that flew this airplane for any length of time knew about this and most had a pair of knee pads in their A-3 bags should they have to make this trip. Unfortunately I was still a FNG or newbie so I didn't have that necessary equipment.

In any event I made my way to the forward compartment trying to make it as easy on my knees as possible while wearing my parachute and carrying my walk-around oxygen bottle. I know it didn't take more than a few minutes but it seemed to me much longer than that. After reaching the forward compartment I strapped in on the instructor's station behind the EW's position and figured that was the end of my duties until I would have to make the trip back to the tail for landing and taxiing.

The mission proceeded as normal though refueling and feet dry over South Vietnam. As we were getting ready to set up for the bombing run we were notified that the aircraft behind us in the cell did not have an operable bombing system so they would not be able to drop their bombs on the selected target. All bombs on target was the holy grail of all of our missions and any crew dropping their bombs outside of "The Box" was sent home. I know of one crew who met that fate while I was there.

Our friendly staff officer asked me if I could go back to the tail and use my radar system to direct the aircraft behind us into position to drop his bombs. Naturally being a young gung-ho gunner that believed in the motto "C'est la Vie" I literally jumped to attention and said "Can do." Actually I don't believe I was really given a choice by our friendly SO. Capt. Duelfer on the other hand asked me if I would volunteer. I might say at this point that throughout all the time I was on his crew he was one of the best A/Cs I had ever flown with. On a later mission he refused to fly a particular aircraft with maintenance issues and was required to immediately present himself to the Wing Commander to explain his actions. He won but I have flown with a few who would have said "yes sir" and taken off with a sick airplane.

But I digress. I donned my equipment and made my way back to the tail, did my duty and "Dealt Charlie a Bad Deal" by using my radar to guide the following aircraft to the proper release point. After that I crawled back forward and settled in to the IE position for the flight back to Guam. By then my knees were talking to me in a very loud way and I had been notified by the SO that he wanted me back there for the landing and taxiing to our parking spot. One of the duties of the Gunners in the tail was as a lookout to clear the aircraft during turns and maneuvers both in the air and on the ground.

As we neared Guam and began our letdown I again made the trip back to the tail. Unfortunately I had just gotten both knees on the seat prior to setting down, raising the seat back and strapping myself in when the aircraft hit the runway pretty hard. When we hit I was pitched forward and ended up with my head where my feet should have been and my feet where my head should have been. I was sure glad I had my helmet on! By the time I finally got situated the way I was supposed to be and got on the intercom the A/C was frantically calling my name and hoping I hadn't fallen out when the gear came down. I let him know that I was bruised but not amused by the landing. I found out later that the SO had changed places with the A/C for the landing after I had disconnected from the intercom.

Needless to say, when we got on the bus I was pretty steamed and the A/C could tell. He guided me to the back of the bus and sat with me instead of next to the SO up front. At debrief the crew got another look at how some officers, and they are few and far between, are more concerned about how they look than how they do. The only thing the SO could find wrong with the entire mission was the way the crew referred to me as "Gunner," "Guns", or on one occasion "George." According to him I was to be referred to as Staff Sergeant or Sergeant Schryer. I could only imagine what those who flew with him as an A/C must have thought of his leadership style.

As they say though, "All's well that ends well," but this was my introduction to the fact that "Murphy" had followed SAC to Southeast Asia and could show up on any flight on any day. I would see him a few more times, as would all the other crews, before I returned to the Land of the Big BX.

Kadena – X-Wind Launch
Nick Maier

The Arc Light crew mission schedule was a perpetual calendar. No matter where or when a crew dropped into any assigned line, they remained with their line schedule until re-deployment stateside. It was finally my crew's turn to press on and rotate to Kadena Air Base, Okinawa. Instead of flying direct on the Young Tiger cattle car, we were scheduled to fly a bombing mission from Guam. On the return leg we would turn left at the Philippines, and then pedal the short two hour and 51 minutes north to Kadena.

That meant another bag drag of all our earthly possessions, stacked into the BUFF crew compartment for the re-positioning trip. It would be a minor inconvenience. The entire crew was jubilant over the prospect of leaving the stuffy, formal environment of the Andersen AFB showroom war. Every visiting fireman and politician in the world was given the VIP tour at Andy, which always included the Arc Light Center and of course, a strike mission briefing. Since the brass never knew who was going to drop in for these dog-and-pony shows, they firmly insisted on strict military courtesies and behavior. That drove combat crews up the wall, since it was very difficult to live their roles as *"Twelve o'clock High"* bomber crew rowdies, if they were expected to salute superiors, shine their boots and wear official military headgear.

The coast of Southeast Asia lies 2,598 miles, five hours and 43 minutes, one mid-air refueling, and two time zones west of Guam. Perched at the edge of the troposphere on our climbing flight path, best range altitude of 39,000 feet, I could finally see the horizon made hazy by the earth being battered by the sea. From that altitude and position, the coast of South Vietnam looked no different than that of northern California. At that same instant the Navigator announced over the bomber cell interplane radio frequency, "Amber cell, we're at Point Bravo!"

"Amber Two!"

"T'ree!"

The copilot's adrenaline brought him back to life and into action. "Bongo, Bongo, this is Amber at Point Bravo, one one four three Zulu, copy?"

Bongo was the call sign for the Ground Control Intercept radar site, which would monitor our bomber cell for the duration of our in-country bombing mission. The GCI controllers would also relay the flight progress of Amber Cell to Blue Chip, the Combat Operations Center located on Tan Son Nhut Air Base, at the northwest city limits of Saigon. After an uneventful weapon release on the target just south of the DMZ, our three naked swept wing black birds, streaked across the darkened coastline, and headed outbound toward civilization. We started a climb in perfect unison, to take advantage of our greatly reduced gross weight, and cruise at an altitude where fuel economy became a critical factor. It was still four and a half hours back to Andersen for Amber Two and Three, and no place to go when they got there, if the weather should suddenly be worse than forecast during the typhoon season, or some catastrophe should close the runways.

It remained strangely quiet on board all three bombers. The navigator finally completed his calculations that Magellan had utilized 500 years ago. Amber One had reached its planned mitosis point in space, to proceed north heading for Kadena Air Base.

"Amber One turning to zero two six degrees. Amber Two you've got the lead. Sayronara," he broadcasted over the interplane frequency.

It was always amazing how invigorated I felt, knowing that we were not returning to Guam that night. In the early days of the endless stream of bombing missions from Andersen, returning crews would be greeted at the Arc Light Center by the Base Exchange mobile snack bar. This traveling ptomaine trailer had been accurately labeled by earlier generations of GIs bent upon self-administered dysentery, as the "Roach Coach." When it became obvious that the strike missions were to remain an unending way of life, the wheels were removed from the Roach Coach, permalizing its position. For the serious eaters and drinkers, picnic tables were placed into position, and decorated with covers of thatched roofing to emphasize the aurora of perpetuity. The area was naturally called "Gilligan's Island," a geographical Freudian slip that revealed the aircrew's comic hopelessness of their predicament. We would not have to partake of the constantly warmed over chili dogs and re-chilled stateside beer tonight. Moving from Guam to Okinawa was considered to be equal to an R&R by the crews and staff. The bombing missions from Kadena were a small price to pay, and simply came as interventions among the wild escapades of Koza City.

We were directed to a rare landing approach to Kadena's runway 23R, because of a frontal passage shift in the normally persistent north-easterly winds, which permanently streaked and bent everything on the island including the low cypress trees. Located twice the distance north of the Equator than Guam, Okinawa also fell victim to cold fronts pulsing off of the nearby Asian continental air mass, producing low scudded clouds and drizzly bone chilling rains. The overcast night sky revealed each major assembly of human habitation below, by the ghostly white glow pushing up through the black clouds. After an eternity, our descending aircraft finally penetrated the murk, and burst under the ceiling with a forward visibility of five miles. This presented a passing view of the brightly-lit towns strung like baroque pearls along Highway One.

The final landing approach carried us over the northern outskirts of the city of Kadena. There was considerable Air Force activity both at Kadena, and at Naha Air Base located on the southern end of the island, all of which had been originally assigned to this Far East theater, as part of the defense forces for Japan and Korea. The buildup of personnel committed in support of the Vietnam conflict had increased the island's military population to post-Korean War levels. The Pentagon never officially acknowledged that there were B-52s at Kadena, since their presence on the island and their use for bombing missions in Southeast Asia had caused serious controversy in the Japanese Diet legislature and among all the home island populace. The native islanders of Okinawa would become the most vehement of the world's anti-Vietnam war protesters. At the slightest provocation, in every city and town on the island, the incensed demonstrators would materialize out of nowhere by the hundreds. As if on cue, they would manifest their furor by snake dancing with locked arms down banner lined streets, each participant wearing a wide white headband inscribed with bright blood red anti-war slogans. To accompany their dance of protest, they choraled a loud rhythmical oriental chant of anti-American hate, conducted by a political activist riding in the rear of a Toyota pick-up truck, equipped with a Sony public address system, demanding an end to the use of Kadena Air Base as a haven for the despicable American B-52 bombers. To us inbound SAC tourist's amazement, the facade of every building in the entire town was ablaze with brilliantly colored red and white banners, splashed with bold black Japanese ideograms, fluttering in the strobing glare of floodlights and huge swaying paper lanterns. When I switched our bomber's landing lights on, the complete approach zone was lined on both sides with 50-foot long pennants, weirdly waving in the night on 10-foot high spindly bamboo poles. Each placard was cluttered with Japanese anti-war slogans. There were enough of them smeared with American taught English "Hated Yankee Go Home", which miraculously translated all the others.

On the narrow highway running just outside the perimeter fence, barely 1,500 feet from the edge of the runway, I spotted the rotating red lights of the local police vehicles. As our huge black aircraft completely filled the scene with sound and presence, I could plainly see the scuffle of a mob being forced by club wielding uniformed police to clear the roadway. Some of the hysterical demonstrators were pressed up against the air base's eight-foot high chain link perimeter fence, which was topped with a pyramid of triple-row barbed wire. They were attempting to use their long poled banners as defense weapons, while others were throwing their shattered poles like spears at our landing bomber.

"Looks like the natives are restless, troops!" I informed the crew. "You don't have to read Japanese to know the base will be closed by curfew tonight. But we're in luck, tonight is Thursday and its Animal Night at the KOOM."

The crew's disappointment cluttered the interphone as I completed the landing roll out, and taxied through the narrow maze of taxiways leading to the fan shaped rows of black steel revetments. Each one was identical by the wonders of American standardization, to those left behind on Guam some 10 and a half hours ago. It will require another full length WWCD story to revive the memories for everyone who has experienced the infamous Animal Night event at the Kadena Officers Open Mess.

The next day it was raining that cold drizzle found only on Okinawa. The SAC briefing room at Kadena was considerably smaller and austere, with none of the glitter and pomp of the Andersen Arc Light Center. The room was a carbon copy of USAF operations around the world. Strike briefings were also less formal than those presented on Guam, since Okinawa was out of the mainstream for junketing politicians and Pentagon brass on their fact finding trips to 'Nam. As a result, the briefing staff was a little more laid back in their ceremony, except of course whenever the wing commander was in attendance. He was a typical SAC, no-nonsense leader, who was interested in only one purpose, get the assigned job done - and prevent the natives from interfering with the missions.

The additional noticeable aspect of these briefings was that the audience here also included the KC-135 tanker crews. They would fly in buddy cell with the bombers, until the air refueling was completed just north of the Philippines, then return to Kadena. As soon as the staff meteorologist started his slide show, it was obvious that there were other factors not usually dealt with flying missions out of Guam. Tonight's weather was going to be grim all the way through air refueling, with numerous thunderstorms associated with a frontal passage, then clear sailing to the target, until our return north back into the large air mass system.

When our formation of three tankers and the mated three bombers finally pulled up at the runway hold line, I wondered how many of the birds were reporting in the green with minor malfunctions being overlooked, rather than attempt repairs in the mini-typhoon conditions battering the ramp. It had been a long time since any of us BUFF drivers were forced to use the crosswind crab system for take-off. This Boeing brainchild was unique among the entire world's aircraft, and would simplify our rapidly approaching take-off challenge.

There had been a constant stream of three-way radio chatter between Charlie, the SAC DO, and the tankers regarding the peak gust factor of the night's winds. Finally, the fearless tanker crews launched, with the maximum crosswind perilously close to their airplane's limit. Since their flying gas station was originally designed as an airliner equipped with conventional landing gear, they had no Rube Goldberg crab system to steer them into the face of the shifting gale. They would rely on the time honored, Orville Wright wing down into the wind, and fight like hell to keep the bloody thing going straight.

"You'll need nine degrees crosswind crab to the left, pilot," my copilot reported after calculating the angle from the tower's wind information.

"Roger, nine left," I repeated as I turned onto the active runway for launch.

When the line-up for the standard rolling take-off was completed, I was conscious of sitting slightly off center due to the crab setting. The blackness of the driving rain was barely punctured by the beam of the sole crosswind landing light, attached to the right forward gear truck. Its sealed beam of searching spotlight was staring down the slick yellow runway centerline, which was almost obliterated with long streaks of rubber left behind by countless landing aircraft. I aimed the gradually accelerating mountain of metal into the awful murk, constantly checking and re-checking the dim red wall of gauges and instruments for any telltale indications of a catastrophic malfunction. Any attempt for an aborted take-off could only be initiated before the aircraft reached final decision speed. Once passed that critical indicated airspeed very few pilots survived the experience of aborting and stopping on the remaining runway, much less trying to explain to an accident board their reasons for even considering beating the astronomical odds against them.

One millisecond past that calculated speed committed us to continue the take-off, and hope that there was sufficient aircraft power available to get safely airborne. The upward ejection seats designed minimum safe altitude and airspeed combination was zero feet and 120 knots, which at least gave the occupants a calculated chance for surviving a bailout while still on the take-off roll. As Captain of his craft, the airplane commander was morally duty bound to stay with his stricken B-52 to at least 500 feet above the ground, for the navigators to survive their downward ejection and the gunner to escape from his large hole in the sky.

The windswept roar of eight, water-augmented J-57-P-29WA Turbojet engines gave credence to the advertised 100,000 pounds total thrust, which forced the nearly invisible black aircraft closer and closer to the threshold of flight. For the crewmembers in the forward compartment, the engine racket was a powerful sonorous pulsation, mixed with the bow wave of rain and wind wildly searching for an entrance into our dry capsule. The gunner was absolutely lost in a tremendous plume of jet engine exhaust and rainwater spray.

Finally, for an infinitesimal nanosecond, 225 tons of Seattle slag became suspended one millimeter above the 12 millimeters of static rainwater, which had accumulated on the two-mile long sea of concrete. This physical combination placed the bomb laden aircraft at the thin edge of uncontrollable hydroplaning, and was the last grip of an uncompromising Mother Nature's attempt to keep the fleeing machine earthbound.

"Coming up on unstick speed - - Now!" The copilot's voice almost broke on interphone, betraying his apprehension over the incredible events going on around him. He was unconsciously listening to the splash of wind and rain on his windows, which could be heard above the din of engines and aircraft systems. With a classic case of white knuckles, he held his own control wheel, attempting to stay in step with my maneuvering of the buffeting controls and chattering rudder pedals.

It required the sum total of my flying experience, to keep the metal encasement straight down the narrow dark strip long enough to break ground. Suddenly we entered another startling but seemingly more secure experience, the unforgiving black void of flight. Take-off airspeed at last, and the airplane literally became unstuck from the planet's gravitational grasp. The BUFF's blunt nose started to rotate up until I cranked a thumb full of nose down stabilizer trim, beginning the elevator riding climb into the clouds out of sight from my last reference to the rain battered surface below, the blurred glow of the runway lights.

"Gear up!" I ordered, as I pressed the rudder pedals to brake the still wildly spinning main gear wheels, before they entered the close confines of their cavernous wells. The steaming rubber was already friction heated dry, and could cause untold damage to critical aircraft systems located in the cramped wheel well spaces, if any of the tires had shredded or tore even the smallest amount.

"Gear up," the copilot echoed as he pulled against his shoulder harness inertial reel, leaning forward and reaching with his left hand to raise the wheel shaped knob to the up position. The gear doors slammed shut, causing an immediate blackout of the landing light glow which reflected off the intimately close black rain cloud. The entire sequence from runway roll out to gear up blackness only accounted for 65 seconds of the Crewdog's entire allotted life span. There were six aircraft launched into that maelstrom that night. The remainder of all our given days would be spent alternating between consciously reliving those horrendous moments, and trying to suppress the resultant sleepless nights.

Guam is Good, but Kadena is MEM-OR-ABLE!

Red 2 - 7 July 1967
Toki Endo

Two events caused me to finally break my silence about my experiences on 7 July 1967. The urging of friends and that of Rod Gabel, the navigator on Red 1 whose own story appeared in an earlier volume of *"We Were Crewdogs."* The second was a picture of George Jones' granddaughter, Mary Galemmo, rubbing the name of George Jones, her grandfather, onto a piece of paper from the Wall. The picture was published in the February 6, 2009, American Journal.

On this mission the target was a suspected Viet Cong supply area along the coast, 45 miles east of Saigon and southwest of Vung Tau. It was to be a MSQ directed bomb run. The cell of three B-52Ds was to fly a VFC formation for MSQ radar-guided bomb run. My crew from Columbus AFB, Mississippi, was assigned aircraft 56-0595 and was designated as Red 2. The crew consisted of Capt George Westbrook, pilot, Capt Dean Thompson, copilot, Capt Chuck Blankenship, radar navigator, 1st Lt George Jones, navigator, MSgt Olen McLaughlin, gunner, and me, Capt Toki R. Endo, the electronic warfare officer. It was a day mission with an early take off so General Crum could return in time for his farewell dinner at the "O" club later that night. After our preflight, we were lounging around in front of the aircraft and I remember George Jones smiling (George seemed to be smiling always) and saying, "Generals always pick milk runs to fly." As it turned out, the mission turned out to be just the opposite.

That was my third TDY to Guam. I previously had a short 15-day, a three-month, and a six-month trip in F-models. It was my first tour in D-models. The crew had been together for about three months. The Columbus AFB and Tuner AFB crews had been merged into one squadron at Columbus AFB. All of the crews were a mixture of Columbus and Turner crewdogs.

We flew the standard route from Guam to the jump off point. The route headed southwest from the Pre-Initial Point (PIP) to the Initial Point (IP) and then had a 110-degree right turn onto the bomb run heading. The bombing altitude was 32,000 feet. Something made me decide to ride this bomb run completely strapped in, chap kit on, helmet and mask in place with the clear visor down. I have no idea why I did on this bomb run. The cell was in the VFC formation prior to the turn at the IP. Red 1 had contacted the MSQ site. The MSQ site reported that it had contact with Red 1. Just prior to starting the right turn, the MSQ site reported it had intermittent contact with Red 1's beacon. During the turn onto the bomb run heading, the MSQ controller directed Red 1 and our jet to switch positions after he determined that our beacon was operating better than Red 1's.

I heard George roger the command and state that he was descending. I heard him say that we were level at thirty-one five. I heard Red 1 acknowledge the call. The copilot of Red 3 was flying the aircraft but relinquished control to his pilot until the cell lead change was completed. Red 3 dropped further away from the formation to give the other two plenty of room to maneuver.

Everything seemed to be going well until I heard someone over interphone call out, "Look out, lookout, we're going to hit!" Then I felt a slight bump and thought we just run into some wake turbulence. A short time later I heard a large "whump" and then was slammed violently to my right and pinned there by a tremendous force to the right side of the ejection seat. I thought we had stalled and were in a spin. I thought the pilots were trying to recover from the spin. It seemed like the pilots were having a tough time trying to recover because the G-force pinning me to the right seemed to last forever. Then I heard a weak voice twice over interphone saying, "We ought to bail out!" I waited for the eject command. Then I felt and heard another "whump!" I immediately thought it must be really bad because the navigator had just ejected and thought I better eject since I'm supposed to be next. I don't recall any calls made by the guys "downstairs", George or Chuck. With the G-forces were pinning me to the right side of the seat and I really had to struggle to grasp and rotate the handle and squeeze the trigger.

Red 3's copilot later reported that he saw Red 1's wing cutting through our 47 section and the tail section go spinning away. Both of Red 3's pilots saw Red 1 roll over on its back into the missing wing side and then disappear from view. They stated that we kept our attitude and continued flying for several seconds before we disappeared in a huge fireball. They saw engines and other pieces flying out of that fireball. They didn't believe that anyone could have lived through that catastrophic explosion.

After what seemed like an eternity, I finally managed to grab the handle and squeeze the trigger. The best way that I can describe the feeling is to imagine being in a centrifuge pulling really heavy G's and trying get your right hand to move down to grab an object The next thing I remembered was a feeling of tumbling through the air, with my mask being yanked to and fro from my face. I felt the parachute opening on my back and then I was swinging below the canopy. I looked around for either the aircraft or other parachutes. It was an eerie, nightmarish scene with a gray pallor to it - like the end of the world. I was between two cloud layers. I remembered seeing sunlight before we had started the turn on the bomb run heading. Now all I could see a grey cloud layer above me and lumpy clouds just below me. I saw multiple plumes of black smoke trailing downward against a gray background in the distance. There were fiberglass panels and what I thought were pieces of the padding/insulation from the cabin area floating downwards. The padding/insulation was burning also. I saw a spinning, burning fiberglass panel knifing towards me almost horizontally. I prayed that it would miss my canopy. It did but it came close. It looked like one of the panels from the bunk area. I didn't remember seeing any fire while we were spinning downward. I guess I was concentrating too hard on reaching the ejection handle and trigger.

There must have been some fire because the medics found several bits of molten aluminum that had hardened and embedded in the cloth of the legs of my flight suit. I also had several spots of reddened skin on my lower legs where I must have been exposed to fire at some time. Later the accident investigators concluded that I was exposed to a fire while still strapped-in. Also, that I managed to eject between 14,000 and 12,000 feet based upon my first conscious feeling after ejecting, the nearness of lower clouds and the reported tops of the lower deck of clouds.

I released my seat pack but stopped short of cutting the four rear risers. That's when I noticed numbness in my right arm. I looked at my right arm and saw a dark stain around the elbow area. My first thought was that it was hydraulic fluid. I knew I had to see what was going on with my elbow. Then a panic thought struck – "Don't look because your arm may be gone!" I quickly suppressed the panic because I knew I had just used that arm and hand to release the survival kit. I rolled back my sleeve and saw a hole in my elbow. I could see white bone. The dark spot was blood. I knew I had to take care of it once I got down.

Shortly after releasing the survival kit, I entered the lower cloud deck. It was like being in a thick misty fog. I bounced around a bit and then I started swinging like a pendulum. Now, I wished I had released the four rear risers. I wrapped my right leg around the line leading to the survival kit and raft and pulled up on it trying to counter the swinging. Somehow it worked.

I broke out of the clouds and saw the water. I thanked God that it was the ocean rather than land after remembering the target was supposed to be a VC supply area. Also, I had lost my chap kit. They never did make the belt and leggings small enough for me. It didn't dawn on me that I was far short of the target area. I saw the brown water where the river water met the ocean. I was going to land just on the blue water side. I saw land beyond that blue-brown line but couldn't tell how far it was to land. If I had to guess, I would estimate that I landed about a mile or so from shore. I saw two other chutes in the distance and it looked like they were going to touch down in the water also. I managed to turn the canopy so I was facing into the wind. I looked down and saw the raft hit the water. I watched my feet until they hit the water. As soon as they hit, I released the canopy fittings. The wind must have carried away the canopy because I didn't have any problems getting tangled in the risers. I pulled the cords to inflate my water wings. The water wings inflated quickly but I should have waited until I got into the raft because it was a struggle to get into the raft with them inflated.

I remembered that I needed to take care of my arm. I hauled in the survival kit and unzipped it. It was full of greenish-yellow water. Water had gotten into the kit. The greenish-yellow color was from the shark repellent or dye marker. I found the first aid kit and treated my elbow by squirting some kind of antiseptic into the hole and wrapping the wound. It wasn't easy doing it with my left hand. The seas got rougher. As I crested every wave I tried to see if I could spot the other two people that I had seen while I was descending or even the shore - no luck. I noticed a streak of the greenish yellow dye-marker/shark repellent trailing behind me and up the wall of the wave. I tried to deploy the sea anchor because the wind was really picking up and I was afraid of being turned broadside to the waves. I found I couldn't undo the wet masking tape that kept the sea anchor cord bundled together because my right-hand wouldn't cooperate and the slipperiness of the tape. I just left the sea anchor with a short lead. It seemed to work because the raft remained pointed into the waves.

I remembered to shut down the survival beacon. I tried the survival radio but it was dead. I had lost my recently-purchased Seiko (BX special) watch. It had a bracelet type band that didn't stay on during the ejection. I had to guess about the time and when I estimated 15 minutes had passed, I turned on the survival beacon for about 30 seconds and then turned it off. I tried the emergency radio from the survival kit but it was dead. If it had worked, I should have heard other emergency beacons. I couldn't hear a thing – not even static. Everything was made tougher because of a right hand that didn't want to seem to cooperate that well.

Sitting in the raft, I remembered the briefing the Intel guys gave us about the VC using junks to move their supplies. I especially remembered the part about how the inspected and safe junks had a blue square with white stripes painted on the deck. A panic moment set – I assembled the survival rifle to repel hostile boarders. That was really a chore. Then, my rational-self came to the rescue and asked, "What can a single-shot 22 hornet do against a boatload of AK-47s?" I dumped the rifle along with the dead survival radio overboard. The Intel guys had also briefed about the VC using survival radios to lure in the Search and Rescue (SAR) teams and then ambushing them. I didn't want that on my conscience.

After what I guessed was about an hour had passed, I spotted a jet low on the horizon. It seemed to be making circles. Every time it pointed its nose towards me, I would turn on the beacon. When he turned away from me, I would turn off the beacon. I don't know if it helped but finally he came toward me and spotted me. It was an F-106. He did a 360 around me and waggled his wings after I waved at him. I returned a "thumbs up" to him. I felt a tremendous relief. He then headed off in another direction. It was a lonely feeling after he left because I didn't really know what would happen next or even when.

It seemed like an eternity before I spotted a helicopter heading toward me. I thought the SAR guys had forgotten me. It was a Marine supply H3. It dropped a horse collar. The crew chief operating the hoist managed to convey to me that they wanted me out of the raft and in the water. I reluctantly followed his instruction because the raft had become my "security blanket." I managed to get the horse collar under my shoulders despite the numbness in my right arm and the water wings. I remembered that they dragged me through the water. I dropped off once because my right arm just wouldn't let me take the dragging. They dropped the horse collar again and I managed to get hoisted aboard. I later found out that they were trying to get away from the raft just in case it was flipped up by the rotor wash and hit rotor blades or so they said. At the time I thought they thought I was VC and were trying to drown me. I thought my tremendous command of every four-letter word in the English language helped convince the crew chief that I wasn't VC. Of course he probably never heard me since he had his helmet on and the racket the H3 was making.

Once aboard I saw the pilot (Major Suther) and copilot (Capt Willie Creeden) of Red 1 sitting on the canvas seats lining the fuselage. The H3 flew to another spot and picked up another person. It was my copilot Dean Thompson. He was barely conscious. I noticed that his right arm had been flayed so that a large part of his muscle and skin was peeled back towards his elbow. There was very little blood oozing was his damaged arm. I guess my Boy Scout training kicked in and I knew I needed to perform some type of first aid. I hollered at the crew chief for a medical kit. He pointed to a small, olive-drab first aid kit snapped to the fuselage. He then resumed looking out of the side door for more survivors. The first-aid kit looked like one of those small aircraft first aid kits. No matter; I found some antiseptic and poured on the exposed muscle and remaining tissue. I wrapped his arm in a gauze bandage to close the wound. Then I noticed Dean going into shock. I dragged Dean back further into the fuselage, away from the wind coming through hatch opening. I put his feet up using some flak jackets and hugged him trying to give him some heat from my body. I thought about that later and realized that it was a stupid move because we were both in soaking wet flight suits.

The H3 dropped Dean off at Vung Tau but the medics wouldn't let me go with him. Major Suther, Willie, and I had to go to the triage center at the Army hospital at Tan Son Nhut Air Base. There, we were poked, prodded and x-rayed by army doctors who had to put up with a SAC ADVON Lt Colonel who yelled at the Army that he didn't know how to treat people who had ejected. He was quieted, thanks to a great Army nurse who threatened him with a huge hypodermic if he didn't vacate the triage section. I remember the doctor unwrapping the gauze that covered the hole in my elbow because when he removed it, it must have pulled away the clot. I apologized for spraying blood on the cubicle curtain. The doctor just nodded and said to the nurse that it will have to be treated later. He then covered it back up with the same dirty covering.

That's when I saw George Westbrook. The medics rolled him in on a gurney. I didn't recognize him until he spoke. His face was swollen like a pumpkin, his left eye was completely closed, the back of his flight suit was still charred and it looked like parts of his harness had melted into the back of his flight suit. It took me several months before I could stand the smell of roast beef because of George's burns. I found out later that a clip board George used to keep on the glare shield flew off and hit him above the left eye. Dean had looked over at George and thought George was dead because of all the blood. When George regained consciousness, he tried calling on the interphone for everyone to bail out. Little did he know until he saw the helmet later that the clipboard also cut his interphone chord. George remembers trying to fly the plane (what was left of it). He remembered that the airspeed indicator was pegged. He didn't eject until he broke out of the cloud deck and saw water. All he remembered was hitting the water and he went so deep that it got cold and black. He managed to inflate his water wings and find the surface. He was too numb to get free of his canopy but managed to struggle into the raft. He remembers there were sharks bumping the raft and seeing their fins occasionally. When the rescue helicopter arrived, he couldn't believe the pararescueman jumped right in with sharks to cut him free of the canopy and put him in a basket. The PJ waited in the raft until they lowered the basket for him.   The accident investigators estimated that George had ejected below 3,000 feet (the reported scud layer at the time) going close to Mach One and with his parachute a streamer because the fire had melted part of the canopy. His recovered parachute was indeed, partially melted. The only thing that saved George that day was the angle at which he hit the water. To me, George was the hero on our crew. He tried to give everyone a chance to eject at the risk of his own life.

At Than Son Nhut, we were finally put in an Air Force medevac ward. I remembered being driven from the Army area to the Air Force side of the base in a school bus with wire mesh covering the window openings. I later found out that the mesh was to prevent hand grenades form being tossed into the bus. We were in the same ward with some young wounded soldiers who were headed for Japan for further surgery. They were in pretty bad shape. I remember thanking God because we were the lucky ones with only minor injuries.

About five hours after landing at Than Son Nhut, I finally got my right arm treated but it was really painful because they had to irrigate the hole to clean out the sewage I had picked up during my time in the water. My right arm had locked up by then and the x-ray technicians had to bend it straight so they could take x-rays of it per a flight surgeon's request. I guess the Army x-ray didn't satisfy the Air Force. The straightening of the arm really smarted. Then they x-rayed my entire body. While they were trying to align me on the table, one medic said, "Did you know your spine is bent to the right?" My comment was, "not before this flight!" I remembered being wheeled to the treatment room in a wheel chair. What struck me was all of the buildings were sand-bagged and that drove home the point that there were people trying to hurt us. That made me very nervous while lying in the medevac ward with an unprotected window across from me. Of courses I had no idea how far we were from the base perimeter.

The flight back on the KC-135 was quiet. All of us were in our own deep thoughts. Back on Guam, we were met, not greeted, by the brass. The only one who really said he was glad to see us was our Squadron Commander, Mo Moran. Mo later told us that the brass wouldn't even let him go out to the plane. However Mo managed to sneak out to meet us by riding with the driver of the truck that had the stairs on it unbeknownst to the brass. However, he was not allowed to accompany us to our quarters. They placed us in a house in the family housing area. Once there, the flight surgeon asked if I needed anything. I said a case of cold Budweiser. By golly, he wrote out a prescription for the case. The shock of what happened and the realization that Chuck, George, and Mac were gone finally hit me after a coupled of cold ones. That night was the first time I that I cried since I was a little kid. To this day I wonder why I was spared. The other three had wives and children. I was a bachelor with only few who would grieve my loss. I still don't understand it and probably never will.

If there were any humorous incidents to this event one was the Army nurse threatening the Air Force Lt Colonel staff weenie with a syringe if he didn't stop hugging them while they were trying to examine us. Then there was the Army nurse who saw us while three of us, Major Suther, Willie Creeden, and I were waiting to be x-rayed. She was a Major who seemed very concerned about what had happened but also asked Major Suther and Willie Creeden if I spoke English. There was also me asking the medic who syringed out the crud in the hole in my elbow if he thought I was crazy to tell him about any more cuts that needed cleansing. He had used sterilized water and then alcohol to clean out the hole in my elbow. Next he applied a liberal dose of, I swear, 100x-strength iodine. I believe they had to pull me down from the ceiling after that treatment. Rats, he found several smaller cuts/abrasions to treat even though I lied through my clenched teeth.

With approximately two months remaining on our TDY, I was returned to flying status. I had an "over-the-shoulder" flight with George Jackson and then Dave Niebauer. After that, I was declared "safe." I was returned to flight status as a spare EW. I didn't get many flights because I got the feeling that crews were reluctant to fly with me because of the memories I brought back to them. It was like I was a pariah. I only flew two more missions during that time. I was relegated to augmenting pen aids for most of my time.

George recovered from his injuries and went onto fly about 60 days later. Dean did as well. Unfortunately for me, part of whatever hit my right elbow left a couple of bone chips in it. The treatment was to let the chips dissolve themselves. It took a Navy orthopedist to convince the flight surgeon not to operate because of the high risk of infection. They immobilized my arm in a sling and took x-rays about every 30 days. After the chips dissolved, my elbow was treated with whirlpool baths to breakdown the scar tissue that had formed. What a great relief to be able to return a salute with my right arm. The commanders made a decision that we would remain on Guam because of the morale problems we might create if we returned to Columbus AFB shortly after the accident as the only survivors of two tragedies. SAC figured that we would have a demoralizing influence on the families. You see the day after our collision; we lost five more crew members when one of our crews coasted off the runway at Da Nang Air Base into an uncharted mine field, caught fire and exploded. The crew had to divert to Da Nang because of complete hydraulic failure.

When we returned to Columbus, I was given the choice of whether I wanted to be crewed with George again. I said "heck yes!" So we crewed together for about four months when George was transferred. I switched to another crew for my deployment in '68. To me, George did everything a pilot should do; he kept trying to give everyone a chance to eject by attempting to control what little was left of Red 2. I know, privately, he took the loss of our three crewmates hard. I guess as the crew commander one cannot help but feel that way, especially when you are flying the jet

George has since passed away. I don't know what happened to Dean Thompson. I do know that he returned to Vietnam in F-4s. In 1997, some partial remains of George Jones and Chuck Blankenship were recovered and identified. Their remains were returned to their families for burial. Unfortunately, Olen McLaughlin remains listed as Missing-In-Action (MIA). When the results of the accident investigation were released, the blame was laid on the feet of the staff for using a formation in violation of the B-52 flight manual. The report also chided the pilots for not over ruling the MSQ-controller's request (it was given as a command and not a request) to change positions during the 110-degree turn.

Additional events –

While in the medevac ward, a Lt Colonel from the SAC AVON team asked us to write down our memories of what happened just prior to the collision. I explained because of the numbness in my right arm, I couldn't write very well. My lower arm was so numb that I could not make my wrist bend. The Colonel said to just jot down phrases and they would make them sentences. I laboriously scratched out my thoughts in very simple phrases omitting such things as verbs. About a year later I was given a copy of the classified accident report to read. Lo and behold, there was a photo-copy of my disjointed, verb-less thoughts just as I had penned them. It read like almost unintelligible disjointed ravings of an idiot. I was thoroughly embarrassed by the almost incomprehensive characters that were sprawled on the page. No where did the report explain my inability. That just reaffirmed my belief that I should never trust a SAC staff weenie!

We Were One Plane Ahead
Jack Hawley

Let me give you a little background here - I never really understood why the personnel assigned to my first crew, were assigned as they were. Generally, or so I thought, a new AC got an experienced copilot, and if the Navigator was new, got an experienced Radar Navigator, with the EW and Gunner having been around awhile. When we were officially designated as a crew, our total experience, adding up the experience of each in their certified positions, totaled less than six months.

These guys were good at keeping their AC out of trouble. In order Bob (CP), Lou (RN), Bill (N), Jim (EW), and Dick (G) were all first class acts. Let's press on. We were flying out of U-Topia (oops, U-Tapao) and remember, these guys were good. I was designated the Air Borne Chimpanzee (ABC) (oops, a little senility setting in) Air Borne Commander, and we were headed for a target in the vicinity of the DMZ. All was smooth IP inbound, and here's where it gets a little foggy - just about release time when the EW says he's acquired SAM lock-on. We, as cell lead, released our weapons, and since we were in the post-target turn, the Gunner states, "Pilot, this is the Gunner...Number Two's been hit." My reply was a calm "Roger." Yeah, you guessed it. Upon realizing what I had actually heard, the tape recorded my comment about five seconds later as a not so calm "Oh shit!!"

As memory serves me, Ken, the AC of the plane hit had lost three engines, 37,800 lbs of fuel in the left drop tank, all kinds of instrumentation and hydraulics, and it was later discovered, that shrapnel beneath both the pilot's and copilot's seat had rendered both ejection seats inoperative. Number Three withheld their bombs. Barry, in the number three position requested guidance. I stated, "Proceed to the secondary target for release." He came back with the book answer (yeah, I forget the reg) says, "I can't proceed to the secondary target as a single aircraft." I replied, "You have my name and authority. Proceed to the secondary target." He did so with successful results.

Meanwhile, the CP, RN, N, EW, and G were keeping me out of trouble, keeping all the mission stuff straight as Number Two flew off our wing and we guided him to DaNang and coordinated his approach and landing. The EW called to say we were ordered to return to base long before I thought we should leave Ken. Ken put the airplane down in one piece at DaNang with the help of some obviously talented controllers.

Upon landing, we did our normal post-flight stuff, and as I headed to the bus, the mobile coordinator (gee was that Foxtrot) said, "Jack, you're coming with me." Like, I was wondering why. Little did I know that a SAC contingent of generals had arrived at U-Tapao to check up on the operation status of the base. Upon entering the Command Post, I was greeted by Maj. Gen Salvador Felices who was not disappointed with what we'd done to help the Number Two flight crew and complete the mission. Methinks there were about another 16 or more stars in that Command Post Conference Room at the time.

Had it not been for five top notch crewmembers, Number Two could have perished. Lest we not forget, the Devine Guidance involved in this entire effort.

I dedicate this story to the Gunner, Dick Sager, who, after retirement, was shot and killed in the convenience store where he worked in north Fort Worth.

Chapter Four

Bar stories [bahr] [stohr-ee] – *noun* - a narration of an incident or a series of events or an example of these that is or may be narrated, as an anecdote, joke, etc. told at a counter or place where beverages, esp. liquors, are served to customers.

A dubious Top Crew poses with Lt. Gen Richard Hoban in 1974. L-R are, seated: Capt. Dave Thomson (radar navigator), Hoban, Capt. Tom Strange (pilot), standing: 1Lt Steve McCutcheon (navigator), SSgt. Don Emerson (gunner), Capt. Charlie Hillebrand (copilot), Capt. Dave Hofstadter (EWO).

How to Do Bad and Still Be the Top Crew
Dave Hofstadter

In the 1970's SAC was, for the most part, divided into two numbered air forces. The 15th Air Force was headquartered at March AFB, California, and generally contained the bomber, tanker, and missile wings in the western US and Pacific. The Midwest and East came under the 2nd AF (now the 8th AF) at Barksdale AFB, Louisiana. Each was commanded by a three-star SAC general. In 1974 the 2nd AF commander was Lt. Gen. Richard Hoban. Every year he held a conference in his headquarters recognizing the top bomber, tanker, and missile crews from each squadron across 2nd AF. In that year we Crewdogs were all back from Southeast Asia and were busy requalifying for SAC's nuclear alert mission.

We of crew E-62 at Carswell AFB, Texas, were no exception. We were busy getting ready for our upcoming crew check ride. In fact, all of us in the 9th Bomb Squadron were so busy training that our squadron commander had not selected a crew to go to General Hoban's upcoming 2nd AF Top Crew Conference. Our pilot, Tom Strange, heard this, and, being ever on the lookout for a boondoggle, suggested to the commander that crew E-62 was available on that week and would be happy to go, right after our check ride. The commander agreed that he should send someone, so we were selected on the spot. Pretty slick. A week of banquets, tours, praises from all the brass and rubbing shoulders with all the hot-shot crews in half of SAC. Just for the asking. Well, on to the check ride.

Soon we were all strapped in a B-52 - the six of us, pilot, copilot, radar navigator, navigator, electronic warfare officer and gunner, and three wing evaluators, one for each rated area for our check ride. I should note that there's not room for any more than that in the B-52D so there was no evaluator for the gunner.

Now, on a SAC check ride, you are supposed to fly a very precise route during a low altitude bomb run, defend the bomber against simulated threats, and electronically demonstrate precise bombing accuracy, all tracked and scored by systems on the ground and the on-board evaluators. That's what's supposed to happen. You are NOT supposed to get disoriented, fly some other route of your choosing, let the bomber get "shot down" by missiles and yell at each other over the interphone. The evaluations were not pretty. Except for the gunner, we were all "busted" - unqualified as crew members. A recheck could not be scheduled before the Top Crew date. The squadron commander faced a dilemma. Should he send a not qualified crew as his Top Crew?

He did. No one down at Top Crew was any the wiser. It worked. We had a great week of wining and dining with all the generals and over 30 other SAC crews. We never let on. In fact, for one of us, it was quite the contrary. At one cocktail event Tom Strange stepped up to Lt Gen Hoban and asked, "Just how DID you make general, General?" General Hoban looked at Tom for a moment while the room went silent, and said, "You know, Strange, not too many years ago I was a wise ass captain just like you."

The Infamous 70-Knot Call
Jack Hawley

This is a true story in remembrance and dedication to Lt Col Laymond N. Ford, a B-52 instructor pilot with an instructional prowess, unequalled by most. Those who knew him, and witnessed his mastery of instruction, had the utmost respect for him. He was a quiet, humble man of God, who treated everyone he knew as a friend and professional. I had the privilege of being his copilot for two of my six Arc Light tours. Had it not been for, at that time, Major Ford, I believe I would not be alive today.

Major Ford was a true believer in equal participation, regarding the takeoff and landing phase, and on one mission, it was this copilot's takeoff. How well we remember the phrase, "Coming up on 70 knots...now!" at which point, the navigator started the stopwatch. As S1 timing was close to expiration the navigator would state, "S1 timing...now!" This copilot, forgot to make the 70-knot call. Unbeknownst to the copilot, the pilot had casually "punched" the instrument panel stop clock, knowing exactly the S1 timing and speed. This copilot, realized the error with the "committed" call by the pilot at the expiration of S1 timing, and this "co-" proceeded to go into the "Oh shit" mode, as he continued the takeoff/departure procedure. The normal climb out checklist was finished, level-off was made, auto-pilot was engaged, and the mike button was keyed by the pilot saying, "Co-this is the pilot." The response was a cautious, "Yes sir!" The pilot quietly questioned, "How many children do I have?" The copilot responded with, "Six, sir!" Major Ford's calm quiet reply was, "I'd like to see them again."

No other comment, no critique, no added words...nothing, was ever said again, regarding that screw-up, on that takeoff, on that day. In his remaining 18 years, that copilot/pilot/instructor pilot never made that mistake again. How interesting, that such a subtle comment could have such a profound effect.

Postscript: Many of the RNs and Ns remember those 'exciting' syllabus 'proficiency sorties' to practice some flight basics and reminders, with the instructor pilot onboard, ie. approach to a stall, etc. Upon completion of syllabus requirements, Maj Ford would rebrief the crew, all three of us, sometimes four, that certain activities would occur, pulling circuit breakers, simulated loss of engine(s), electrical power, and others.

Now as an aircraft commander, one night at return 'flight level' for decent into U-Tapao the gear was lowered. We experienced close to total electrical power failure, rendering most lighting and flight instruments inoperative. I used needle-ball and the row of lights on the ground at U-Tapao as a backup attitude indicator, while the electrical problem was partially fixed, enough to land several minutes later. Had it not been for Maj Ford's added training incentives, that flight would not have gone as smoothly.

B-52D Ferry Flight to Guam (Yawn)
Bill Reynolds

Recently I read one of the official monographs on Linebacker II. That was an interesting time not experienced by many of us who later flew the B-52D. The stories from combat missions over Vietnam, particularly over the North, are deservedly famous as are the crewmembers that flew them. Later, for the BUFF Crewdog it was lots of alert and being qualified in conventional and "non-conventional" (whatever that means) missions. But sometimes we had to do things like take one of our aircraft to another place and return with a different one. This story is about one such flight and fits the definition of aviation well: "Flying is hours of boredom, punctuated by moments of stark terror."

The full crew collaborated in writing this story. The pilot/aircraft commander was Bill Stiller (Lt Col, USAF, Ret.). The copilot was Bill George (Col, USAF, Ret.). I (Bill Reynolds, Maj, USAF, Ret.) was the radar navigator. Mark Chapin (Col, USAF, Ret.) was the navigator. Tony Adamcik (Col, USAF, Ret.) was the electronic warfare officer. I think SSgt Dennis Harmon was our crew gunner during that time, but he did not make this flight. No one recalls a gunner being on board. Also on board was Lt Col Vern Prather (from the Wing Bomb/Nav shop), flying only as an "observer" and not in an instructor capacity.

188

Each member of the crew contributed his particular memories to this account. Except for Lt Col Prather, we were all assigned to the 20th Bomb Squadron, Carswell AFB. Note: a B-52H and its crew were lost in July 2008 off Guam. It was a 20th Bomb Squadron crew from Barksdale AFB, Louisiana. It was daytime, good weather, and they were not flying a purely training mission.

A ferry flight is a flight where a plane is moved from one place to another. It is just a transfer - nothing more. Swapping out tail numbers between Carswell AFB, Texas, and Andersen AFB, Guam, is not a mission from which we would expect anything more than a lot of flight time (36.6 hours, in this case). According to Flight Records, the aircraft we flew to Guam was tail number 085. The total flight time was 16.4 hours (16.0 of that in daylight) on 10 August 1979. Now, more than 29 years later, each member of the crew can recall it, some with remarkable detail. It is interesting that as we worked on writing this story, each person's recollection would improve with time, and he would remember more details.

We took off from Carswell short of our planned fuel to keep our weight down due to the heat (Fort Worth in mid-August). Our first air refueling was over California, where we took on approximately 80,000 pounds of fuel. We refueled again somewhere near Hawaii. Tony Adamcik recalls his constant communication with the Hickham AFB Command Post over the status of the tanker. As Tony recalls it, "We hoped they would not show up so we could Remain over Night (RON) in Hawaii." A good TDY was a rarity for the Cold War BUFF Crewdog, and for some reason, aircraft often had maintenance problems after landing in Hawaii.

Although we took on 120,000 pounds of fuel, the pilot requested additional fuel because he knew we were facing headwinds of 140 knots. I don't know how after 29 years the pilots recall those numbers and not getting extra fuel, but they do. Bill George remembers the tanker guys complaining that they were bucking the same headwinds and couldn't spare us any extra gas, but he distinctly remembers Bill Stiller talking them out of about 2,500 pounds extra. That was a drop in the bucket. We burned about 20,000 pounds per hour at cruise altitude, thus the extra fuel gave us only a few more minutes of flight time. At that point in the mission he also recalls being about 10,000 pounds below the fuel curve because the 140-knot head winds were much stronger than the 50-knots we had planned. At the time, that was a SAC rule-of-thumb for flight planning. Fuel planning is very important. One of the aspects of that task was to plan to have extra fuel (reserve) upon arrival at your destination just in case one must divert to an alternate destination. That alternate must be suitable for landing a B-52. At least, that was the plan, but nothing is both close to Guam and suitable. As the saying goes, "An airplane will probably fly a little bit over-gross, but it sure won't fly without fuel."

If you compare the navigation equipment of the D-model in the 1970s to today's aircraft, the difference is astonishing. In fact, the last new USAF-procured aircraft with a navigator position was the B-1, essentially a 1970's airplane. Until the mid '70s, navigators were trained in T-29 aircraft that only had limited rudimentary instruments, few of which remain in use today. Navigators of today (now called Combat Systems Officers) would consider the D-model phenomenally archaic with no INS, GPS, or Moving Map Display. That may be why they no longer even call them navigators. Navigator training, now at Randolph AFB, Texas, (planned to move to Pensacola NAS, Florida, in 2010) has not conducted training in the art and science of navigation by the sun, moon, and stars for over 10 years! Well, enough about the young "button-pushers" of today.

In the Old Dog, we had Doppler, radar, and celestial navigation. Tony Adamcik recalls using the sextant often for celestial navigation, as well as triple-checking our pre-comps. The pilots had radio aids such as TACAN and VOR. However, in the middle of the Pacific Ocean, the only useful navigation aids were Doppler and celestial. We could measure our actual altitude and compare it with pressure altitude for what is known as "pressure pattern navigation." The BUFF was also equipped with navigation dead-reckoning analog computers, but these were totally dependent on accurate airspeed, drift, and groundspeed inputs from the Doppler, which often has "issues" with the smooth surfaces of the ocean water.

The flight to Guam was during the day almost the entire way. The many position reports were Tony's job, and he still remembers how "funky" the HF radios got around dusk. So, we had the sun, a questionable Doppler, and pressure pattern navigation. Our skills at low-level radar bombing were well honed and we considered ourselves quite good at it. However, our limited faith in sun lines, pressure measurements (we seldom did in training) was of questionable value, and a less-than-precise Doppler made the long over-water adventure a challenge for the navigation team. But we managed. We found the Hawaiian Islands and the tanker and got our gas. At that point about half of the "adventure" was over. It was boring, but everything seemed fine, except for the fuel situation.

We were also experienced (if you can say that about lieutenants and junior captains) in navigating around in the central USA, but not over the ocean. We had experience in the avoidance of thunderstorms by "20-nautical miles." The weather between Hawaii and Guam on 10 August 1979 included many thunderstorms that we either avoided or over-flew. However, zigzagging to dodge thunderstorms meant flying farther, and much stronger than predicted headwinds also caused us to fly longer and thus use more fuel.

As the time passed and we burned off fuel, we became lighter and climbed to higher altitudes to be more fuel-efficient, ending up at 40,000 feet. So, "there we were" at Flight Level four-zero-zero, dodging thunderstorms and heading for Guam. As we approached Guam, it was dark! There were thunderstorms literally everywhere - forget 20 miles. We were just trying not to fly into one of those things. As I directed the aircraft through a break between two large radar returns, I found out that all was not going to go well for us that night. We ended up boxed-in and surrounded by thunderstorms and lightning. No matter where I looked, it was just more "walls" of storms.

When we found our island, we were still at 40,000 feet and directly overhead. We did not need all that altitude, but "speed is life, altitude is life insurance." So, we were well-insured. All of that flying around to avoid thunderstorms plus headwinds tripled our planning factors and ended up costing us one of our most valuable commodities: fuel. We were going to have to land very soon. We no longer had fuel to divert to a suitable field nor could Andersen AFB launch a tanker for us because of the same thunderstorms. According to Bill George, we could have made one or two approaches, but then our fuel supply would be depleted. Island destination fuel reserves of 50,000 pounds were required by regulation. We got there with just over 21,000, and minimum fuel in the BUFF was 20,000 pounds! To say the least, the pilot crew was more than "just a little concerned." We didn't even have enough fuel to divert. We were praying that nothing would go wrong

191

with the aircraft during the descent, approach, and landing phase. Our fuel level was at the critical point.

I advised the pilot of the problem with the storms and he said, "Well, just pick the softest looking one, and we will spiral down." So we did. The B-52 has equipment to keep ice off of critical areas. But it is very cold at high altitude and very wet in the midst of Mother Nature's mid-latitude storms. The formula was: cold + wet = ice. Our aircraft was covered with it, but we did not know that at the time. Bill George recalls, "During the descent we had so much Saint Elmo's fire coming into the cockpit and all over the windows and front instrument panel, it was very freaky. It was raining so hard that it was like someone was shooting a fire hose at the windscreen." He says that a couple of times he looked out his side windows to check to see if the anti-icing was keeping the engine inlets clear of ice. However, he couldn't see them. The rain was too heavy. He remembers being amazed that the engines were not flaming out with all that water going into them and recalls, "I had never seen anything like it before or since - just incredible!" Needless to say it was very difficult for the crew to stay focused on the job at hand: flying the aircraft, running checklists, navigating, and keeping aircraft systems running properly.

Usually during high-altitude cruise I flew with my ejection D-ring stowed, no helmet, on headset, and quite comfortable (if that is possible in a B-52). At some point during that descent, I donned my helmet and pulled my seat straps tight - very tight! I was seriously thinking that we might end up ejecting at high-altitude, at night, in thunderstorms, over the shark-infested Mariana Trench (very deep). So, when Mark Chapin (navigator) looked at me, he seemed alarmed and proceeded to cinch up his ejection seat straps even tighter. That was one of those moments in life we never forget. Expressions can say so much more than words.

We needed to land. Andersen AFB had been struck by lightning that took out their electrical power (Hello "Murphy's Law"). That meant there were no runway lights, no navigation aids, nothing. We frequently trained doing practice airborne radar directed approaches (ARDA), and we did a real one on that approach. Thinking back to that night, it is amazing how calm we really were. Denial must be a strength of the young (and maybe the stupid). Fortunately, the runway of Andersen Air Force Base on Guam is easy to locate on radar, even if you have never seen it before. I seem to recall Lt Col Prather standing behind me during the approach. I don't know if he was checking on the Nav team or getting close to an ejection hatch!

About two miles out from the end of the runway, the pilot announced that because his windshield wiper had malfunctioned, he could not see the runway. Bill George's rendition of what happened next goes: "When Bill Stiller said, 'Co, my windshield wiper just broke and I can't see anything - you got it!' I was thinking 'you've got to be kidding me!' I haven't touched the stick in hours and all of a sudden you're giving me the jet in the most critical phase of flight. You really must be having problems if you are trusting me to land it in this mess!' At that point I looked over at him and his windshield wiper and realized that he really was serious. Of course all those thoughts and actions only took an instant, and I answered the call appropriately. Even if I sounded cool and calm to the other crewmembers, I was very concerned with the whole situation. After all, that was my first landing at Guam, my first landing in the middle of a thunderstorm, and my first landing at night when there were no approach lights. How was I supposed to get that thing on the ground without having the normal night visual cues to go by? The flashing lightning was destroying my night vision, but it was also giving me an occasional, unexpected, visual cue. Still, the runway environment was all very blurry from the quantity of rain on the windows. Consequently, without really knowing exactly where I was in relation to the end of the runway, I made an educated guess by listening very intently to the Radar Navigator's voice telling me where he thought we were. I initiated all the normal control inputs for a landing. Fortunately our aircraft landing lights were working properly, so as we got closer to that black hole of a runway the approach end and centerline markings were illuminated - we were lined up dead center of the runway. When I saw the runway markings I continued to pull the throttles all the way to idle and kept trimming nose up and pulling back on the stick and just waited for the touchdown.

Normally, in the BUFF, when you pull the power off for landing, after a few seconds you feel the crunch of the wheels hitting the runway. If the time is excessively long it usually means you pulled the power off too high above the runway, and you are about to have the bottom drop out from under you as the airspeed bleeds off, resulting in a bone-shattering, plane-bending, hard landing. Some B-52 pilots may consider this "normal." On that landing, it was to be one of those, or so we thought. The sound of the engines had diminished long before, and none of us had yet felt the runway. It seemed like the bottom was going to fall out at any second. But it had not, and the longer it went, the worse the outcome was going to be. After an inordinate amount of time, I remember Mark Chapin yelling out 'Are we down?' Not knowing myself, but hoping we were, I touched the brakes, and the whole crew yelled out a "Whoop" of relief as we all felt that sweet deceleration. Evidently there was so much water on the runway and a lucky soft landing cushioned our touchdown, so we did not feel the runway under our wheels.

When I was an instructor at Castle AFB later in my career, I told this story to my students many times. As it turned out, the Guam runway we landed on has a downhill slope to it in the landing zone, so in reality, to have all of our wheels touch at the same time we were actually pointing down when we touched. On a normal, flat runway we would have hit the nose wheel first which would have caused us to porpoise and to have a very bad landing." Another old phrase goes, "Flying is the second greatest thrill known to man....Landing is the first!"

Bill George had landed the monster and everyone breathed a huge sigh of "any landing you walk away from...." We were all relieved, happy, thankful, and probably very lucky. Any number of things could have happened. We had a lot going on, including "a beautiful light show" of Saint Elmo's fire on the aircraft and in the cockpit. At any time, we could have been stuck by lightning or encountered wind shear. We were both lucky and good that night. After we unloaded our aircraft (it is pure rumor that crews from Texas transported Coors beer to Guam), I turned back for one last look. The huge black and camouflage BUFF was still covered in ice, now melting fast in the warm air of the tropical island. What a sight! For certain, we all felt that, "It's better to be down here wishing you were up there, than up there wishing you were down here."

A few days later we took off from Guam to reverse the process. The first attempt at a return flight in aircraft 092 was on 15 August and lasted 4.4 hours to burn down fuel to return to Guam and land because of a fire light on #8 engine that illuminated after we committed for take off. The pilot recalls Guam air traffic control cleared us for a 250-mile radius around Guam from 1,000 to 5,000 feet to burn down fuel to a safe landing gross aircraft weight back at Andersen. Tony Adamcik recalls the clearance as "to fly aimlessly about." Every one on the crew got some stick time, and we did an aerial tour of Tinian, Saipan, and other Northern Mariana Islands.

The following day, we flew aircraft 092 to March AFB, landing after 13 hours (seven of that at night) because we did not have sufficient fuel to fly to Carswell. We then made the short 2.8-hour flight to Carswell, landing and ending the adventure. According to the flight records, we landed at Carswell sometime after midnight on August 17th. We didn't crash, nobody died, and it was all over in the last few minutes of a very long day.

Nobody got any medals. The only history of this event is the one each of us has in his memory. But the memory is one of those "Crewdog" things that still binds us together after almost 30 years. But the truth is that each crewmember knows that we probably owe our lives to the people who trained us and made certain that we knew and could do the things that we might need to do some day. We had a beer, talked about it, got some sleep, and turned around to do it all again. Oh, yeah, we were Crewdogs! About a year later I was assigned PCS to Guam for two years. But that is another story. Did I ever tell you about the time...?

The Longest Crawl
Jack Cotrel

My career as a B-52 gunner really got its start in the summer of 1969. I was serving a one-year PCS assignment with the 8th Tactical Fighter Wing at Ubon, RTAFB, Thailand. At that point of my AF career my specialty was "Aerospace Ground Equipment Repairman". It was not the most pleasant job on a 100+ degree flight line. Then things took a turn for the better when the NCO assigned as an admin clerk in the shop's Administrative Office happened to ask me if I could type. (Thanks Dad for making me take that semester of typing.) Shortly thereafter I was off the flight line and in a nice air conditioned office. I had the responsibility of keeping the statistics, typing correspondence, and various other "duties assigned."

One day when sorting the shop mail, I came across a flyer sent out by SAC recruiting "B-52 Fire Control Operators and KC-135 Boom Operators." I did not hesitate to choose the B-52 option. Being a B-52 crewmember seemed like a much more exciting life. I rethought that decision a few times later when we were evading SAMs. With a little help from my shop OIC and NCOIC, my application was soon on its way. Within weeks I had "Flight Orders" assigning me to the 51st Bomb Squadron at Seymour Johnson AFB, in Goldsboro, North

Carolina. Note: Seymour Johnson was the only stateside base I was ever assigned to for permanent duty. The orders also had a class assignment at Castle AFB, California, for Gunnery School and there I was 24 months later flying combat missions over Vietnam.

My wife and I were married in September of 1970, a couple of weeks after I finished gunnery school. We set up housekeeping and endured the ORI's, and the Alert Pad until December of 1971, then the Arc Light orders came through. This is just a little background on how the following Crewdog tale came about.

I'm sure we all remember how emergency procedures (EP) were hammered into us during our Crewdog days and with good reason. One of those EP's was specific to gunners on the B-52D. That was making a crawl through from the gunner's compartment to the flight deck in front. I recall thinking that this will never happen to me. Well, it can and it did.

I believe that it was in July of 1972 that our crew was scheduled to rotate back home for our 28-day break. Whoopee! Our crew was made up with the following personnel: Major Dick Rinehart, AC; Captain Jim Simms, CP; RN, (unknown); Captain Steve Geiger, Navigator; Captain George Nelson, EW; and me, SSgt. Jack Cotrel, Gunner. Somehow we were selected to ferry a D-model from U-Tapao to a base in Texas. I believe it was Dyess. Naturally, we were anxious to get home and it was my choice to ride the tail, although it seems like Major Rinehart invited me to ride up front and I declined. Bad decision.

As I recall, we took off from U-Tapao at 0h-dark-30 and all was normal i.e. hotter than Hades. I had been sweating profusely and had my compartment air conditioning cranked to the max. As we approached cruising altitude, I figured I was cool enough and turned on the heat. It did not take me long to figure out that I had a problem. It was not getting any warmer. It gets pretty darn cold very quick at 35,000 feet, evidenced by the condensation freezing on my windows. I seem to recall consulting the TO's and trying to get things working but no dice. My only choice was to go forward or risk being a Popsicle. After informing the boss of my situation, I could have sworn someone said, "Aw hell, let him freeze."

Major Rinehart obtained the necessary clearances to descend to 12,000 feet and it was time to make the crawl. Mind you, I don't think this was ever practiced by me, just read about. I cannot recall putting on my chute, but given the extremely confined nature of the crawlway, I don't think I did. I opened the hatch at the rear of the gunner's compartment and thought "Oh Crap! What have I got myself into?" The crawlway was around 12 to 18 inches wide, and I remember thinking "Here I am in one of the largest aircraft ever built and I am going to have to duck-walk 50-plus yards to the front." Even though I did have my helmet on, the noise was unreal. I recall that there were stations to plug my helmet into along the way so I could give progress reports to Major Rinehart. I'm pretty sure he was anxious to get back to altitude due to fuel considerations. We had a tanker rendezvous that would be difficult to reschedule. I remember at one station he asked me what the hell was taking me so long. I repressed the answer I wanted to give and politely said I would hurry. Fat Chance! There was no such thing as getting in a hurry on that catwalk. Between very sharp protrusions and very hot surfaces, caution was the prudent course.

I remember when I reached the bomb bay and called in, and some smartass said "Open bomb bay doors." I'm pretty sure it was my nav. Seeing those massive tires in the retracted position was somewhat humbling. I remember thinking that if that gear goes down I am in big trouble. After what seemed like another mile, I could see my nav up ahead looking out of the electronics bay which was right behind the RN and Nav stations. I remember that I was exhausted when I finally reached my destination. Again our crew had its share of comedians and one of them told me I forgot to shut my hatch and would have to go back and shut it. Thank the Lord he was trying to be funny. I would rather have bailed out than do that again.

We arrived at Dyess AFB in the early afternoon. We were all dead on our feet. I recall one of the crew chiefs inviting me to dinner at the NCO Club. I told him I would take a short nap and join him. I woke up at 5:30 AM. We got into Seymour Johnson late that afternoon where I met my son for the first time.

While those days were hard, I would not trade the experiences and the friendships made for anything. You guys know what I am trying to say.

I dedicate this "tale from the tail" to MSgt. Walter Ferguson, who perished in Charcoal 1, the first B-52G shot down in Linebacker II. It happened near Yen Vien on December 18, 1972. Fergie always had a kind word. And also, of course, to all the other Crewdogs who have made the ultimate sacrifice.

The BUFF and the Soviet Trawler
Dave Hofstadter

So, it's the fall of 1973, and the war in Southeast Asia is drawing down and the number of combat sorties is dropping off. But there is still a whole ramp full of B-52s on Andersen AFB, Guam. So SAC puts a bunch of us on alert - with nukes. That was something several of us had yet to experience. Still there are BUFFs and crews to spare, so the brass says, "In between your combat missions, let's go fly some training missions. Idle hands, you know." So that's what we do.

There were no training ranges out in the middle of the Pacific, so we just planned a mission to fly around some of the dozens and dozens of little volcanic islands in the Mariana chain. Maybe we'd take the opportunity to practice some real low level flying. In fact, on the day this story took place we were popping up and over those lush green island-mountains, so close to them, and the ocean, that we are often very close to (read that as "well below") minimum downward ejection altitude for the nav team on the lower level of the BUFF. But what the heck; it was clear-and-a-million visibility. So up we went over a tall island, and we could just see the peak of the next island as we approached the crest of that one. As we prepared for the weightless "yahoo" as we went over the top, let us leave our BUFF there for a minute, aircraft nose high, and navigators throwing up.

We all knew that the Soviets were allies of the North Vietnamese and supplied them with intel. And we all knew the Soviets stationed "fishing trawlers" out there in international waters to collect data on our bombing operations and informed the North Vietnamese about what was coming. Way out there. Nowhere near the Marianas. Wrong!

Let's return to our BUFF as it crests the height of the island and starts down the other side to skim the smooth water between islands. Not so fast, flyboys! Directly in our path was a huge gray "fishing trawler" flying a big hammer and cycle Soviet flag, bristling with antennas, and with decks full of "fishermen" in white Soviet uniforms. Our ECM gear lit up. We had surprised them. We had surprised ourselves. We jerked the BUFF back into a climb - more throttles, and more throwing up.

I also saw throwing up aboard the ship as we passed just barely mast-high over the "fishermen."

After that, we called off our roller coastering over the islands for the rest of the tour.

Buy-the-Beer Marks on the Nose Gear
Dave Hofstadter

B-52 landing gear are lowered from the fuselage, under the nose under the aft fuselage. Now those of you that flew in Southeast Asia may remember that the left-most front tire usually had strange markings around the left sidewall. These were handwritten abbreviations for the six crew positions in the BUFF written with white grease pencil. We carried white grease pencils to write on checklists and ECM scopes and write greetings to Ho Chi Minh and Jane Fonda on the bombs. So, evenly spaced around the sidewall of that tire was written P/CP/RN/N/EW/G, standing for pilot, copilot, radar navigator, navigator, electronic warfare officer and gunner.

Now when the BUFF was parked after a mission, one of those crew positions would end up at the bottom of the tire, where the rubber met the tarmac. The custom was that the crewmember who's position mark was on the bottom would buy the beer after the mission. Statisticians will try to tell you that over time the beer buying will even out among the crew. But this was not true within the B-52D fleet. There was a clear—even overwhelming—occurrence of the copilots buying beer far more often than all other crewmembers combined. How could this be? Copilots immediately thought that the crew chiefs were in on some scam by signaling the pilot somehow, but this was never found to be happening.

There were many secrets passed down among BUFF crewmembers and one was the on-board parking assist found only on the B-52D. You see the D-model had bombing optics—a bomb sight through the floor, that had evolved from the Norden Bombsight. It was long before replaced in use by the radar and largely ignored, but it was still functional down beside the radar navigator. It was offset enough to allow the radar navigator to see the sidewall of the left front tire. So, as the BUFF was rolling to a stop, a prearranged signal—a click on the mike, an interphone comment, or some prearranged signal would have the pilot stop with "CP" neatly and exactly on the bottom.

Many was the time our copilot would be the first out of the hatch and race around to that front tire only to throw up his hands in surprise, saying, "ME AGAIN?" Copilots usually figured this out (or someone squealed) just in time for a new copilot to rotate onto the crew from the States, and more free beer.

My Collection of BUFF Tales
Tom Jones

Museum-Piece Alert

As an aircraft commander at Carswell in early 1983, I was pulling alert with my D-model crew when, during an engine start exercise, our alert aircraft experienced a failed alternator. The D's alternators were bleed-air driven by four individual turbines, using superheated air piped in from the compressor sections of the eight engines. Our particular brand of Thompson alternators were in short supply, because the D was being retired later that year.

Maintenance advised us that our bird would have to come off alert; no spare alternators were available, and none were available to cannibalize off another mission-ready aircraft. We were resigned to having to swap airplanes, with all the overhead that entailed for maintenance, munitions, and us, when some innovative troop remembered where he could find a spare D - with Thompson alternators. That B-52 had already been put out to pasture as a display airplane, but it was close by: the bird was sitting in the little airpark across the runway from Carswell, next to the mammoth General Dynamics factory (accompanied by a B-36, B-58, and T-33, and other forlorn retirees). The maintainers soon got permission to visit the static display B-52 and extract a functioning Thompson alternator. Our bird was fully repaired about a day later, without ever having to surrender its alert line. Clearly, though, we were scraping the bottom of the barrel for D-related maintenance items.

In October of 1983, most of the D's were flown to Davis-Monthan's boneyard, while a lucky dozen or two wound up at static display or museum locations across the country. More than 25 years later, I still enjoy visiting those aircraft, most of which I flew at least once. I recall visiting the Pima Air & Space Museum in Tucson in late fall of 1983, surprised to find that the D-model they had just put on museum display was the very tail number I'd flown for my AC check ride two years earlier. My plane was a museum piece, and I was 28 years old.

Look Ma, No Hands!

My aircraft commander on the 20th BMS crew E-14 was Captain Benny Jubela, who was also a capable IP. We often had an extra pilot along for our sorties, ready to jump in the left or right seat to accomplish some currency approach or landing type while in the traffic pattern at the end of the mission. Once we taxied back after a typical 11-hour-plus mission to pick up the 7th Bomb Wing's commander, a not-particularly-popular colonel.

He jumped into the left seat with aplomb to notch a few touch-and-go's to keep his currency, with Benny riding right seat, and me, the regular co, in the jump seat. On our first trip around the pattern, we knew what we were dealing with. Coming in on short final at about 200 feet above the ground, the wing king pulled all eight throttles to idle and proceeded with his "power-off" landing technique. The results were predictable. He misjudged altitude in the flare and wound up out of airspeed and ideas about the same time. I braced myself on the back of the pilots' ejection seats. Benny grabbed the yoke as we descended rapidly to the runway, sucking in enough back pressure to cushion the blow. WHAM! We struck on all four trucks simultaneously, Benny ramming in power and getting the nose up so we could avoid a second crash-landing. The bewildered nav team downstairs later told us they were scowling and laughing simultaneously.

The colonel tried the same technique next time around, but Benny was ready with a firm hand on the yoke and a fistful of throttle to cushion the touchdown. The third pattern was also dicey – the colonel liked two hands on the yoke, none on the throttles, leaving Benny to feed in power to keep us above stall speed or grab a foot of back pressure to cushion our plummeting touchdowns. After his three patterns, the colonel was good enough to yield: "Thanks for the work, boys!" He smiled in self-satisfaction and headed downstairs to jawbone with the nav team. So, our wing commander wasn't the best pilot – but at least he had many other flaws.

Comm Check

Our crew had a new EW, joining us for his very first stint on alert. We had eight bomber alert lines at Carswell in 1982, so on any given week on alert we had many friends with us, all up for a good practical joke. Before heading out for our pre-flight on Friday morning, I caught up with our copilot and one from another crew to set things up.

Before the new EW called in from his flight station on UHF to do his daily check in and authentication with command post, the two copilots switched their #2 UHF radios away from Channel 9 to a discrete common frequency. The EW called command post with our sortie number, and the usual request, "Radio check." Back came the voice of our neighboring EW, armed with the proper code group, "Loud and clear, Sortie 4. Authenticate Alpha, India." This of course required our new EW to reply with the proper authentication letter. "Authenticate Bravo," he answered.

Silence for a moment. Then the pseudo-command post answered, "Negative, Sortie 4. Authenticate Tango Zulu." Our EW replied hurriedly – but wrong again. No matter what correct code letter the rookie E-Dub gave, our colleagues in the BUFF next door would always give him a verbal thumbs down.

Both our crews were grinning like idiots at this good fun, while I could hear the EW desperately consulting with the nav team to find his mistake. Finally, after a fourth failed attempt, our jokesters next door radioed, "Negative, Sortie 4. Have your AC and you report to the command post after this morning's briefing. Command Post, out." Our EW came scrambling forward, a desperate, plaintive look on his face. "Tom, look at the code book! I've rechecked the date four times! I know I've been giving the right response. Look!

It was all I could do to keep a straight face. "It's OK, EW. We'll figure this out later. Don't worry about it. We wound up back in the shack with our poor EW, at wit's end, dreading the trek to the command post for a dressing down. When we had most of our two crews together in the nav team's room, we listened to his explanation one more time. Then his counterpart from our neighboring crew spoke: "Sortie 4, authenticate Kilo Whiskey." As he recognized the voice, the slow dawn of realization broke over our rookie's face. He took it like a man. Gotcha! Far from being angry, he laughed as hard as the rest of us – he was looking forward to his turn as "command post."

St. Elmo's Fire

I didn't fly combat in the B-52D – too young for that, you old Crewdogs! – but during my five years with the 20 BMS at Carswell I had many memorable flying experiences. On night training missions across the central U.S., we would often cruise at FL 390 or so in high, thin cirrus. One moonless night, the clouds a translucent soup of ice crystals outside our windows, I watched, fascinated, as glowing blue tendrils of light spread from our windshield wipers across the base of our front windows. Soon, jagged tentacles of neon-blue snapped and danced across the window panes, while static electricity snapped and popped in our headsets over the UHF. Looking out my left wing, the same filaments of blue sparked from the rims of each of the four engine nacelles, the St. Elmo's Fire flickering along the entire length of our wing. I'd read about the phenomenon in accounts by sailors on tall-masted windjammers, but had never seen it in the air before that night flight in a BUFF. I saw the fire on several more sorties, and the aurora (far to the north, beyond Montana) as well, but that first magical exposure to St. Elmo reminded me of the special world we airmen inhabited during our long, high-altitude hours, alone at the top of the troposphere.

Airmanship

Now I wasn't there, but one of my favorite stories of B-52 airmanship centered on the exploits of my 1977 Academy classmate, Terry Young, an easy-going, slow talking pilot from Oklahoma. On the ground he piloted a vintage black El Camino, but he was just as at home in the cockpit of a black-bottomed D-model.

Shortly after takeoff from Carswell in 1983, Terry and crew were still on runway heading at 180 knots when the airplane's nose pitched up dramatically. Terry and his copilot shoved the yokes forward, but even at full power, airspeed began to bleed off. A stall and loss of control was imminent. He couldn't pull power to lower the nose, because every knot of airspeed was critical, but they had to get the nose down. The yokes were jammed – they wouldn't move full forward – but Terry found the electric trim button on his left yoke handle was still effective. By running the trim full-nose down, he arrested the pitch-up, and over agonizing seconds got the nose headed back toward level flight.

Declaring an emergency with Ft. Worth departure, Terry and his copilot next remembered the nose-down pitching moment induced by flap retraction. Carefully, they milked up the flaps, a few degrees at a time, accelerating and giving them a comfortable margin over stall. Now they had even more authority, with the trim button driving the all-moving, hydraulically driven horizontal stabilizer, to keep the BUFF's nose level.

With the airplane stable, but the elevator (and yoke) still jammed in pitch, Terry consulted with Carswell instructors on SAC's Command Post frequency. Soon he was tied in to maintenance experts from Boeing and the Air Force test community. The plane had ample fuel to remain in the air for hours, so with elevator control for a flaps-down landing questionable, Terry and crew headed for Edwards and the miles-long runways on the dry lake bed there. The plan was to execute a no-flap landing on the lake bed, keeping enough speed to assure pitch control using only stabilizer trim.

Burning down fuel to landing weight, Terry and his crew had ample time to rehearse their approach procedures. At Edwards, he flew a series of no-flap low approaches, each controllable, at successively lower airspeeds, until he felt comfortable keeping up with pitch inputs using trim only. His last approach brought him straight in on a flat, steadily descending trajectory, requiring no landing flare before touchdown. As he rolled the D-model onto the lakebed, his crew reported the touchdown as barely perceptible. "You greased it!" shouted the Edwards test force commander to Terry; it was a near-perfect landing.

Maintenance investigation discovered that the push-rod driving the air-driven control tab on the elevator had broken on take-off, piercing the elevator sheet metal and jamming the elevator in nearly the full-up position. Only Terry's quick thinking and flying skills, along with his crew's professionalism and superb teamwork in dealing with the emergency, enabled the crew to avoid loss-of-control and a dicey low-altitude bailout. Terry, I believe, received the Distinguished Flying Cross, while his crewmates each received the Air Medal. Well done!

Fuel Leak – Global Shield

We were completing our air refueling during the first leg of our wing's Global Shield Exercise in 1982. A couple of dozen D's were ahead of us in the stream as we buttoned up our A/R doors and resumed our cruise to the low level range. From up ahead came word over the Command Post frequency that one of our fellow crews had a serious problem.

Behind the tanker, the bomber's refueling manifold ruptured, just behind the cockpit and forward of the EW station. JP-4 poured into the bunk area and flooded downstairs around the nav team before the crew disconnected and cut the fuel flow. Their crew compartment was saturated in jet fuel, and fumes filled the air.

The crew went on oxygen, declared an emergency, and took the cabin pressurization switch to Ram/Dump to vent the crew cabin, simultaneously starting a descent to 10,000 feet. By minimizing any switch throws or avoiding any relays which might throw off a spark, the anxious crew burned down fuel and managed to recover at Carswell. It was a lost sortie, but one that successfully avoided becoming an inflight fireball.

Low Level at Red Flag

W.W. was my AC as we bored in toward the target at Red Flag, somewhere in the Nevada desert. The day was brilliantly clear and hot, and turbulence muscled our D-model around as we tried to stay below threat coverage at 400 knots true and 400 feet off the ground, the designated terrain avoidance altitude.

W.W. took our bucking BUFF lower, clearing the ridge lines by a couple of hundred feet – a single wingspan – and sledding down the reverse slope to cross the desert salt-pans at minimum altitude. The trouble was, the AC's definition of "minimum" – forged from his long experience as a forward air controller in O-2s – was different than mine. Hustling across the dry lake beds, I watched the radar altimeter unwind: 400...300...200...we were just a wingspan away from impact. Down we went, until the radar altimeter needle was bouncing around at 150 feet or less. This was too low for me, especially for an exercise, but I didn't want to argue about it over the intercom. Instead I shot the AC a pleading look and jerked my left thumb upward over the throttle quadrant. Take us up! Grinning and obviously enjoying the sleigh ride, W.W. ignored his too-cautious copilot. He dropped us to 200 feet or less every time we bottomed out in a valley. A 20-degree bank would have put one of our 3,000-gallon tip tanks in the dirt.

No fighter was going to get in a shot at us from below, and any aggressor would have some real trouble getting missile lock on us against the ground clutter – although we were very close to becoming part of that clutter ourselves! Somehow we survived, and the RN even managed to get enough peeks at his radar during ridge crossings to find the target. I'm not sure if the tail gunner's eyes were as wide as mine as we skittered across the desert, but I know that I never want to be that low, that fast, again.

Hard-Boiled Egg Strike

Those three-hour celestial navigation legs were long. Everyone was worn out after low level, yet there we were at FL 370 or 390, driving across the heartland of America for sun or star shots. The EW poked the sextant periscope up through the port above the BUFF's bunk area; in the D, the bunk mats were just forward of his tiny forward-facing crew station at the rear of the flight deck.

I looked back from my copilot seat between sextant shots one day to see the EW smiling conspiratorially as he peeled the hard-boiled egg from his box lunch. Instead of gobbling it down, he looked directly at me, then fed the egg, pointy end first, into the open sextant port. There was a loud thwup! and a momentary rush of air – and the egg was gone, sucked out through the port and into the wild blue in one gulp by the pressure differential between the cabin and the rarified atmosphere outside.

Success was all in the technique. If you didn't feed the egg in exactly centered (so my expert E-Dub told me), it could (and did, I can attest) splatter ineffectually down around the operator in a shower of white and yellow fragments. One honed one's technique by collecting ammunition from other crewmembers who were indifferent toward the fate of their hard-boiled eggs. As for me, I liked to watch.

We suspect that egg-ejecting competitions ended when yolk stains started showing up around the ram-air scoop at the base of the 40-foot-tall black, vertical stabilizer. Maintenance had no appreciation for contests of athletic skill and coordination.

Auto-Throttles

A veteran crew always enjoyed introducing a new copilot to his position by demonstrating the automatic throttle capabilities of the B-52D. The AC would demonstrate at cruise, on autopilot, by telling the copilot that some D's had voice-commanded throttle control. To the surprised copilot, who'd never heard of this in CCTS, the AC would nod seriously and proceed to show him the correct procedure. "Engage auto-throttles," the pilot would intone robotically into his mask. "Number Three, idle…now!"

To the copilot's amazement, the #3 throttle lever would glide aft, the tach needle unwinding dutifully to idle. "Number Three, military…now!" commanded the AC, and the ivory knob would glide purposefully forward to the firewall. A good AC might let his copilot practice a bit on his own (coaching him with the right commands, of course) to demonstrate the WonderPlane they were privileged to fly.

Downstairs, the nav team could only hope that their copilot would be amazed at this demonstration long enough to tell some other copilot, or better yet a group of pilots, about his discovery. Only then would the truth emerge: the nav team, listening on intercom, had been tugging the throttle cables back and forth, on command, from downstairs, crouched in that electronics rat's nest behind the radome and beneath the flight deck. Boeing had produced an amazing and fascinatingly capable airplane. Auto-throttles – on a 1956-vintage airframe. What would they think of next?

MITO

Minimum interval takeoff, or MITO, was designed to get the maximum number of aircraft airborne in the minimum amount of time. Strategic Air Command crews practiced it frequently, knowing that once Soviet missiles were launched at our base, we'd have only minutes to save as many bombers as possible for the nuclear counterpunch. The takeoff interval between tankers was 15 seconds, and just 12 seconds for bombers.

The challenge for pilots was to avoid as much as possible the wake of the bomber we followed off the runway. Whatever we did, taking off 12 seconds behind another B-52 meant a very rough ride for the first few seconds in the air.

Taxiing for this exercise amid a cloud of acrid gray smoke from black powder starter cartridges and the exhaust from 80 jet engines on the Carswell Air Force Base ramp, I conned my lumbering B-52D down the short taxiway stem to the active runway. Scrambling to complete the checklist as we approached the hammerhead (where the taxiway met the runway approach end), my copilot and I watched the bomber ahead wheel onto the active. We already had our takeoff clearance from the tower.

"Smoke!" I called when black exhaust spurted from the leader's eight engines, now spooling up to full throttle. Downstairs, our nav started his stopwatch and began counting down 12 seconds. I didn't bother braking as we rolled onto the active, engulfed in the roar and boiling soot from the BUFF in front, now totally obscured in the murk. I peered over the nose, lined up on the white centerline stripe, and, as the navigator's count hit "five," smoothly advanced my handful of eight throttles to the firewall. The big jet shuddered under 48 tons of thrust, shoving us down the rubber-slicked runway. I snatched a quick glance down at the water injection pumps for our eight Pratt & Whitney J57-P-19W turbojets - lights out, water on! Sprayed into the compressor section of each J57, the deionized water increased the mass flow through the engine and jumped the thrust from 10,500 to 12,100 pounds each.

Through the smoke I could make out the white dashes of the runway centerline rolling faster and faster under the nose. The wings shook; rudder pedals juddered as turbulence from the leader swirled and smacked our plane. "70 knots, now!" I called on the intercom. The nav hacked his stopwatch and began counting upward to S1 time; if our airspeed wasn't above the pre-calculated minimum at his call, we'd abort.

My eyes flicked to the airspeed indicator drum, oscillating under the turbulence pummeling the pitot tubes. "S1 speed....now!" called the nav, and a glance confirmed we were above the required 120 knots (140 mph). "Committed," I barked into my oxygen mask, and drew my right hand from the throttles as the copilot slid his palm behind them, keeping the levers firewalled. Both hands now on the jerking yoke, I peered into the gray-brown exhaust ahead and saw the lead bomber levitate into clear sky above the cloud, his eight engines spewing smoke. His wingtip vortices slapped sharply at our wings, and I pumped the yoke left and right to keep the tip gears from ramming the concrete racing beneath us.

For a full minute after applying takeoff power, we accelerated down the runway. I heard the copilot: "Coming up on unstick speed.... Now!" The quivering airspeed indicator showed 152 knots as I pulled smoothly back on the yoke, then popped it sharply to the right to catch the drooping left wingtip, caught in the turbulence.

With the back pressure, the Stratofortress rose magically, the last thousand feet of the runway disappearing beneath our black-bottomed wings. Their tips bent upward with the effort of hauling 450,000 pounds off the concrete. Even as we climbed, the aircraft maintained its disconcerting nose-down attitude, tail jutting upward into clear air.

As I called "Gear up," I rolled right through the turbulence to a fan heading 10 degrees right of the leader's, trying to clear his wake; I could see him above and just right of our track. With a lurch, our big bomber caught lead's tip vortex, the right wing plummeting. The swiftness of the roll surprised me even as I slammed the yoke left to arrest the unwanted bank. For a full five seconds the right wing hung there, seemingly scraping the ground, until full spoilers and aileron dragged it slowly back to level. The air smoothed as we accelerated in a slight climb to 180 knots.

The gunner's voice crackled in from the tail: "Three's airborne." Each second put another football field between us and Carswell - potential ground zero. "Flaps," I nodded to the copilot. The engines dropped a note as water injection thrust ran out, and in smooth air at last I could relax a little. But on an actual wartime scramble, a successful MITO would be just the first of many hurdles on a long and uncertain mission.

When the Lights Went Out
Jack Hawley

Once again, Maj Laymond N. Ford's name, sits at the forefront of me still being alive today. He always gave that extra training to save you're butt, when required. This experience is a tad foggy, as I forget the actual sequence of events on descent but here goes….

So there we were, FL430 methinks, headed back west to U-Tapao Air Base just prior to midnight (dark as a welldigger's, uh, no further definition required). Yeah, we had the descent checklist all ready to go, and the "E-Dub" had already called in the maintenance brevity codes. Hey, short mission, piece of cake, a cool "brewskee" only an hour or so away and here we go with the descent.

Co reads, "Gear handle – Down", roger "Gear Handle Down - OOPS!"

Now the cockpit (and the entire inside of the airplane), looked the same as it did outside, black as a welldigger's; you know the drill. Complete AC power failure…or so it seemed. All we had was the needle, ball, and #2 UHF radio, which at the time, was tuned to Charlie frequency, Ch 11.

"Uh, Charlie tower, we have a problem up here…Upon placing the gear handle down, we lost all AC power."

(To this day, I think we must have had one alternator still on the line; I don't recollect confirming that with the copilot)

I was blessed with a sharp crew. They had already started shutting down all their high power equipment.

Devine intervention again entered the picture.

The U-Tapao parking area (ballpark) lights were an excellent attitude indicator, to maintain wings level.

As we were in a smooth descent with tons of altitude, the copilot carefully attempted and succeeded in getting three of the four alternators back online, with the fourth bus tie breaker closed.

We then had indications of all gear down and locked, proceeded with the appropriate checklist, lowered the flaps a bit early so we got no last minute surprises, advised Charlie tower, and were cleared to land.

I regret that I forget the name of the officer in Charlie tower that night, but remember succinctly, that he was one of the "good guys" who sincerely cared and was always there to help those in trouble, or those with a problem.

I met him in the O'Club bar that night where he asked, "Hey Jack, what the hell happened up there?"

I relayed the problem.

He asked, "How in the hell did you get that airplane on the ground?"

I stated, "As a Copilot, I had an extraordinary instructor pilot, Major Laymond Ford, who taught me well."

He said something like, "Yeah, I know him; you need not explain any further."

A Summer Lake Cruise
Russell Greer

As the S-01 gunner in the 2nd Bomb Wing I scheduled and flew many check rides over the years. Some (most) ended up with the gunner receiving a Qual 1, but there was the occasional Qualification Level 3 that required the individual to receive additional training. That could be ground or flight instruction and then be rescheduled for the check ride. In one case the gunner had received the Qual 3 for a very poorly performed modes check and his inability to determine the correct optimum mode of operation. After the appropriate re-training he was rescheduled for his recheck. Mission planning went smoothly and for his flight we had one of the Squadron Commanders (Will not say if it was the 62nd or 596th) along as AC. The flight was a typical profile of T/O, Air Refueling, Low Level, (Arizona in July), cruise home and about 1.5 hrs back in the pattern for duration of about 10.5. I will mention here that the Stan-Eval branch had just recently acquired a VHS camcorder that we were all eager to use in-flight so it was packed and ready to go for the flight.

Pre-flight went smoothly until we experienced a BNS (Bomb/NAV System) problem that required a bag-drag after engine start. These were so much fun in the 90 degree, 90% humidity environment that we often experienced in Shreveport, Louisiana. This crew was up to the task and the drag and subsequent pre-flight; start, taxi and take-off went off without a hitch. After take-off the gunner began his modes check and I will say that he performed this one just as he had been taught in CCTS and as Donny Ford had re-trained him on over the last two weeks. In other words we had no problems and I was more than happy to grade this as a Qual-1 and leave him alone for the rest of the flight so I moved forward to the IP seat.

Air refueling went smoothly as did the cruise and decent to low level. We were met with the typical mid-summer thermals, which resulted in the normal BUFF jarring mid-summer ride. I will say that the AC was really getting into some aggressive Terrain Avoidance (T/A) flying which honestly tends to be a blast for those of the crew fortunate enough to have a view of the outside (Pilot/Copilot and IP).

One of the highlights of the Low Level Route was Lake Powell. Now typically we had always just flown over the lake, which is situated, with-in some fairly steep canyon walls but the A/C had something else in mind that time. As we came out over the lake he started a decent into the canyon and we leveled off at about 150 feet on the radar altimeter, which placed us about 100 feet below the top of the canyon. That altitude provided a spectacular view of the boat traffic on the lake on this beautiful summer day and no doubt gave the boaters and skiers a sight they were not used to, 200 tons of camouflaged, smoke belching Air Force muscle! Through it all I was running the video camera and trying to take it all in. The A/C was flying and maintaining his distance from the right canyon wall, which was working out fine until we reached about mid way down the lake where the canyon and lake took a jog to the right. That would not have presented a problem IF anyone looking outside had been paying attention to what was in front of the plane and not just what was to the right of it, fortunately the copilot glanced forward and through the viewfinder of the camcorder I saw him begin to motion with his arms as if he is trying to jump. Not knowing what in the world he is doing I scanned to the left and all I saw out the windscreen was canyon wall! At the same time the A/C looked forward, grasped the seriousness of the situation and started hauling back on the yoke and shoving all eight GO FAST levers forward. I dropped the camcorder and along with the copilot see no way we are going to clear the rim of the canyon. As I blindly reaching down with my right hand for the bailout alarm switch (couldn't take my eyes off the impending collision) my hand found the copilot's left hand headed the same way. At the same time we saw the radar altimeter on the EVS screen bottom out as we slid over the rim by no more than a few feet.

All was quite on the mighty BUFF as the A/C leveled off at 500 feet and told the co, you have the plane. After about 15 seconds the navigator called out the time to the next turn along with the heading and it suddenly dawned on me that the offensive team had been silent for some time. It was probably no more than 30 to 40 seconds, it just seemed longer. The rest of the route was completed with very little extraneous interphone chatter. After we climbed out I made my way downstairs and asked the RN to switch to private on interphone. I questioned him as to what he had seen on radar and he stated that he was having trouble making out any terrain while we were in the canyon and noticed that we had a problem at the very last minute, just as the AC started the climb and too late to say anything.

After landing and debrief I made my way back to the Stan/Eval offices but since it was after 7pm no one was left. When I got home that night I popped the tape into the VCR and started reviewing what all I had captured. I had some great video of the pre-flight, gunner's mode check, the DCE and A/R. I also had some awesome footage of the low level and flight over the lake. The last footage of the low level was as I scanned to the left when the Copilot started his arm motions. I had a brief shot of the radar altimeter starting to bottom out as the camera was quickly lowered. The next morning I walked in to see the Chief of Stan/Eval and together we reviewed the tape. At that point he took the tape and headed off to see the Wing Commander. About 45 minutes later I was summoned to Wing Headquarters to brief the 2nd BMW commander on the previous days sortie.

I was never told exactly what action was taken against the aircraft commander but I do know he did not fly any during the next 30 days and his next flight was with Stan Eval. With all of this being said I will state that I had flown with him on many occasions and he was an exceptional pilot and this was an isolated incident, but one in which luck intervened to allow us to fly another day.

Russell Greer
806th BMW (P)
RAF Fairford
Desert Storm
Feb. 1991

I Volunteered For This?
Russell Greer

When Iraq invaded Kuwait I was the Curriculum Development Manager for B-52 Gunners in the 436STS at Carswell AFB, TX. I had qualified in the H-model when I arrived at Carswell but I still maintained my G-model qualification through monthly TDYs to the various G units as well as my sorties with the 20th BMS and 9th BMS on the H at Carswell. Now the trips were not just to maintain currency but I used these to review the new training programs that we were sending to the field with the wing staff as well as to train the wing instructors on the programs, what we now call "Training The Trainer". I was surrounded in the gunnery section of the 436 by some of SACs finest Bulldogs, Scott Smith, Chris Austin, Ken "BO" Regard, Harley Gomes, John Wing, and Pete Gertz. I also had the pleasure of teaching Scott Smith why the ASG-15 FCS on the Gs was much superior to the ASG-21 FCS on the H-model (hey, that is my opinion after over 3,000 hours between the two), in other words I was able to get Scott qualified in the G.

When Desert Shield kicked off all I could think about was the fact that I had trained for years for this and here I was stuck in Texas! So naturally I started making phone calls to anyone at a G unit that I thought could help me get in on the potential upcoming action. Also being the completely unselfish person that I am, I was also trying to get Scott (recently G qualified) and Chris (had flown Gs at Barksdale) in on the action.

Well it worked, sort of. Scott and I were notified that we would deploy to Jeddah as part of that Provisional Bomb Wing so we had to get our EWO bags packed for some hot weather. I left the office on Friday ready to depart Monday for Saudi Arabia. Well as these things have a tendency to not always go off as planned I received a call Saturday informing me that I had been requested by the 97th BMW DO (a former Squadron Commander of mine) to join them when they headed to RAF Fairford on Monday. A pilot from the 436th and I would be picked up by a KC-135 on Monday and travel to England via Pease AFB. I spent the weekend with my parents who had flown in from Alabama and sort of overlooked the need to re-pack the old EWO bag. That would come back to haunt me in a few days. Scott and Chris ended up heading to Jeddah a few days later.

I had deployed to RAF Fairford back in the Eighties as part of the 2nd BMW during BUSY BREWER exercises in the May, June time frame. This trip was taking place in the late January/February timeframe. As we stepped off of the 135 in England I noticed that there was a chill in the air. No scratch that - it was downright cold! Fairford had been closed for about a year when we arrived so billeting was non-existent, plus the fact that we arrived after approximately 2,000 medical personnel that had taken over the BAQ/BOQ facilities. They were deployed there in preparation for the massive casualties that were expected when the war kicked off. Our first night was sleep-where-you-can, so Tex Ritter (CEVG Gunner) and I plopped our cots down in one of the racquetball courts at the gym. The gym also hosted the three working showers on that side of the base. The court had unobstructed airflow from the outside through a large opening that at one time held an exhaust or ventilation fan.

Now that would have been more than welcome during the summer but as our frozen water bottles attested to when we awoke the next morning it was defiantly a formula for getting sick. With the time change I woke up at about 3am that first morning and decided to call home. The only working pay phone at that time was down by the main gate about ½ mile away. It was not a bad walk except that when I went outside there was now about five inches of snow on the ground and it was COLD!!! It turned out that we had arrived into what would be England's coldest winter spell in about 40 years, or so we were told. Luckily even though my bag was packed for Saudi I was able to wrangle a heavy flight jacket to replace the summer one I came with. It did start warming up in a couple of days so my suffering was fairly short lived.

As we on the staff began preparing for our first sorties the Wing Commander notified the medical personnel that they had about 24 hours to find other accommodations since he would not have his aircrews residing in racquetball courts and offices while they waited around for possible action living in billeting. After three nights Tex and I moved into a billeting room with a TV, phone, bathroom and kitchenette - what a difference from our first digs. We actually started inviting the crew chiefs and other maintenance troops to the billeting rooms to allow them to take a hot shower and watch TV. Our first two days worth of sorties went off without any major problems and then I was up for my first mission. Since I was not normally assigned to the 97th BMW and with a large part of their crew force had deployed to Diego Garcia, I ended up with a pick-up crew, 4X? My AC had qualified in the BUFF about a year earlier from 135s and my EW was just about fresh out of Castle. We were lead in a three-ship hitting an airfield Northwest of Baghdad.

Now as a wee lad of six or seven growing up in the 60's my dad allowed me to stay up late at night (9pm I believe) to watch my favorite television show, *12 O'clock High*. I grew up with a love of B-17s and the 8th Air Force and there I was 20-something years later sitting in a morning brief with the 8th Air Force in England, flying a Boeing Bomber into combat. How cool was that? This was all in addition to reading an article back about 1971 in one of my dad's flying magazines (I still have it) about B-52s in combat in Vietnam. Upon seeing a photo from the cockpit of one BUFF looking over the pilot to the lead BUFF I told my dad that one day I was going to fly B-52s. Destiny, fate, whatever, there I was and it is what we had all trained for.

The total duration was scheduled to be a little over 19 hours and everything about this mission went well to start with. The run in to the target was proceeding according to plan and I had the rest of the cell on my scope the entire way in. The pilots were calling out SAM launches but most appeared to be fired without guidance since we were not seeing much in the way of uplinks in the defense station. Intel had suggested the possibility of IR Seeker Heads on some SA-2s. At about 60 TG the AC called out a SAM launch at about 10 o'clock and a few seconds later stated it appeared to be guiding on us. He started a turn to the right and then stated, "Still guiding" even though we had no indications on the warning receivers. At that time we broke left and it still appeared to guide according to the AC. I still had #2 and #3 on the scope at that time and the bomb run continued. The AC called for a right break and still it came. Finally the AC called for a left break and directed the copilot to dive at it. As we entered a rapid descent the EW and I still did not see any uplink indications and the RN was in the final countdown to release. When the co called out "I think this one has us" I stowed the gunnery column and at that time the EW dispensed flares. We heard the detonation after "Bombs Away" occurred in our B-52 Dive Bomber.

At that time the lights were still on, the engines were still singing away and we appeared to be in one piece. It was then that the pilot requested the copilot to help him in recovering from this high-speed dive. I do not know when aircraft control was transferred. We had entered the dive at bomb run airspeed and the throttles had not been retarded prior to then. As the power was reduced and the G-loading increased as they pulled out my next concern was how much altitude had we lost and would the wings remain attached? As we leveled off we heard some different noises outside which prompted the Nav to question what was going on. It seems we had lost enough altitude so that we were within range of the AAA. We were starting a climb back to altitude as well as executing the post target turn to the left when the EW sings out with uplinks at 2 O'clock and began countermeasures, which thankfully proved effective.

Once we reached altitude and slid back into the number one slot we settled down for the long ride home. During the debrief the Intel officer continued to insist that we could not have encountered an SA-2 with an IR Seeker Head but the AC finally told him that maybe he needed to tag along and then maybe he could see for himself. As I walked back into our room at about 9:30am Tex was getting ready for his first sortie that night and asked how it had gone. I looked at him and explained to him that it was sinking in that there were people in this world that wanted to kill us and that we were not on a training sortie through the STRC.

The rest of my sorties were more or less routine with very little opposition. We hit fuel depots and opened the way through minefields for the start of the ground offensive. I will have to say that I would not trade any of the Desert Storm experience for anything. It was what we all had trained to do for many years and as it turned out Scott and Chris did not receive a very warn reception from the deployed gunners at Jeddah. I on the other hand was treated like one of the guys by the great group from Eaker.

E-06
Barksdale AFB, 1985

Standing: Bill Williams (RN), Eric Single (CP), John Burgess (AC)
Frnt: Tony Smith (EW), Russell Greer (G), Mike Ray (NAV), Mike Murray (EW)

Anybody Want a Pachinko Ball
Jack Hawley

As we sat fidgeting, awaiting the start of the preflight brief on Guam, I noticed that Ted (a treasured personal friend) and his crew were the "ramp tramps" that day. They were basically ensuring all was going well with the mission aircraft and the spares, and ready for action should they be needed. Our primary mission aircraft went belly up, thus mandating the use of the spare aircraft. Ted was an outstanding pilot; as was his crew. All were true professionals, never leaving even the smallest stone unturned. When they said the aircraft was ready...so be it.

This particular aircraft had just returned from Okinawa, with the return flight crew participating in the purchase of numerous pachinko machines. As luck would have it, the crew chief was also one who double-checked everything, including the 47 section. He came upon a couple of hundred pachinko balls back there that the return crew had failed to properly secure. Our confidence level was high that we would end with a mission "as briefed." Well, we got pretty close to it, anyway.

Upon our return...

"Pilot...Nav, we're 200 miles out" (of Guam).

"Co...EW, can I have Button 11 in #2 Radio?" "Rodger E-dub."

The EW made the standard introductory call to the command post; however, instead of a request for maintenance codes, the next transmission went something like:

"Blue 1, have you been using your stabilizer trim?" (Duh!...like just for the grins, we'd turned off the stab trim and shut down a couple engines just to make the return trip an "attention getter.")

So I chimed in, "Roger!"

The next response was, "Turn off the stabilizer trim and trim manually...those weren't pachinko balls."

Decision time...enter "Devine intervention", to see us through the next 30 minutes.

Let's see...the stab trim had worked perfectly for the past 13 hours, with absolutely no hint of a problem, and we're looking at a tired flight crew.

I briefed the copilot that I'd be making the landing, with the stab trim on. We would complete the before landing checklist a bit early; and if the slightest problem was noted, the copilot would turn the stab trim off on my command. If needed I'd manually trim for the remainder of the landing sequence, to include a smooth, controlled go-around, if necessary. None of that was required.

After landing, the crew chief plugged in and asked, "Did you have any trouble with the stab trim?"

I said, "Not a bit," to which he said, "Those pachinko balls were really ball-bearings from the stab trim assembly."

Upon deplaning, the crew chief advised us that the DCM had seen to it that every maintenance man on duty was issued one of those ball bearings to carry in their pockets - as a grim reminder of what could happen when erroneous conclusions are accepted. When the crew chief rechecked the 47 section after we'd left for maintenance de-brief, we were advised later he'd found several hundred more.

Crewdog Short Tales
Sam Roberts

First Alert

I had just finished CCTS at Castle AFB, California, and was assigned to KI Sawyer AFB, Michigan. The year was 1968 and I had just become combat qualified as an EW on the B-52. I had not yet been on alert because I had not been assigned to a crew. While at home, I received a call that one of the EWs on a Stan/Eval crew on alert was sick, and I was summoned to take his place.

I packed my bags and drove to the alert parking area where I met him and took over his duty. I went to the alert shack to sign in and determine what room I would be assigned. As I was talking to the airman on duty, the horn went off. "Oh xxxx!" We had to look on the board to determine to what truck and which crew I was assigned. I ran to the crew truck and got in the back seat in the middle. As the rest of the crew arrived, they said, "Who is this guy?" I said, "I'm your EW because your EW went home sick." What a way to meet your fellow crewmembers!

Due to my disciplined SAC crew training, I was able to authenticate the message and perform my duties. We taxied to the end of the runway and returned to the alert pad. What a way to start your first SAC alert! I could hardly sleep for the rest of this first alert tour – thinking the horn would go off again.

Alert Auto Repair

Alert duty at K.I. Sawyer AFB was normally quite boring, and each individual had various ways to stay occupied. I had decided it would be a good time to tune up my car. I walked to the parking area and figured if the alert horn went off that I could easily get to the aircraft by running the 1/8-mile or so. As luck would have it, in the middle of the tune-up the horn went off. I showed my badge to the guard and hot-footed it toward my aircraft. I could see the crew bearing down on the aircraft long before me. They entered the aircraft and, unbeknownst to me, waited until I passed just beneath the engine nacelle which housed the cartridge start (#4 or #6 I believe). At that precise moment the copilot fired the cartridge, and I was engulfed in black smoke and an intense sulfur smell. I immediately hit the ground and visibility went to zero. After the smoke cleared, I got up and went to the hatch to enter. As I did, I could hear the laughter as I came up the ladder. Yep, I had been had!

Alert BX Shopping Trip

It was a beautiful day at K.I. Sawyer AFB. It was probably 40 to 50 degrees. My bomber crew decided to visit the BX on the other side of the base from the alert area. When we arrived there and started to park in the assigned alert area, a tanker crew, who was positioned at the other end of the runway, was already parked in the area with their red license plate in full view. The bomber crew trucks had blue license plates to tell them apart.

We all piled out leaving the key in the ignition and proceeded to shop. Yes, you guessed it, the horn goes off. We run out just in time to see the tanker crew drive off with our truck with the blue license plate. No problem, we will take their truck. Except we all piled into the truck and there was no key! Now what? We only had a few minutes to respond to our aircraft or we would be in big trouble.

We saw a dependent with kids drive up in a station wagon, and yes, we commandeered her car. We told her the situation, helped her out of the car along with the kids, and took her car. I can still see the kids waving but don't recall that she was that happy about the situation. Now it gets interesting because we have to cut across the runway with this lady's car. By now we have security police bearing down on us because they don't know what is happening. We arrive at the plane and respond to the alert as normal. I don't remember how she got her car back, but we were actually given a commendation for responding under such an adverse situation. As per SAC, the local procedure was changed so that all crews were required to leave the keys in the vehicle.

Orange Knife

Remember that switchblade orange knife that was tied to your left thigh? Mine was attached with a piece of parachute cord. I believe the purpose was to cut specific risers on your parachute (if you could remember under duress) to give you control on your decent to the ground. This is my story of the orange knife.

We were a B-52 crew from K.I. Sawyer AFB on our 179-day TDY to U-Tapao, Thailand. We had just completed a combat mission over South Vietnam and had returned to U-Tapao. The bus arrived for our pickup and we had just piled out with our stuff and boarded the bus for our trip to debriefing. The pilot noticed that we did not have the Radar Navigator. We waited for a short while and he asked me (the EW) to go check on the Radar Navigator. As I approached the exit, I noticed his feet on the stairs but he was not moving. I asked if he was okay and he responded that his LPU had inflated and he was trapped (wedged) in the doorway exit. He also said he couldn't breathe. One look up and I could see the LPU culprit. I took out my trusty orange knife and popped the LPU. Thank goodness that was the one and only time that I had to use that knife.

The Gorilla
Ron Poland

I was stationed at Barksdale AFB in the late Seventies when the Pittsburgh Steelers and the Huston Oilers had a fierce rivalry for the AFC Central Division. At that time the Steelers had fan clubs for individual players like "Franco's Army" for Franco Harris. The kicker for the Steelers that season was Roy Gerela and his following was named "Gerela's Gorillas." They, of course, had a mascot in a gorilla suit that could be seen at all the Steelers Games.

The Steelers were scheduled to play the Oilers in Houston, and it was expected to be one of the best games of the season. One of the copilots on a 62nd Bombardment Squadron crew was a rabid Pittsburgh Steelers fan, and he had been able to get tickets to the game. He was scheduled to pull alert duty the week prior to the game so he'd get off of alert on Thursday and would have the weekend off to be able to drive to the game. Unfortunately, the annual Operational Readiness Inspection (ORI) was also expected to hit Barksdale at that time. We were given instructions to stay close to home and be ready for a recall at anytime.

Unbeknownst to everyone on alert that week, the copilot had rented a gorilla suit for his trip to the Steelers' game and had brought the suit to the alert facility. On the weekends, both bombers and tankers aircrews, would assemble in the briefing room for the morning standup. We'd receive the customary weather briefing, and then the senior aircraft commander on alert would brief any hot items or important announcements to all the crews. On that memorable Saturday morning, during the aircraft commander's briefing, this copilot (in his full gorilla suit) came crashing through the door next to the briefing platform and grabbed the aircraft commander, scaring the S--- out of him. Everyone howled with laughter while the aircraft commander took a few minutes to recover from his near heart attack. That was the talk of the alert facility for the rest of the week.

When Thursday came, everyone was warned again to stay close to home and to be ready for the ORI. Sure enough, on Friday we got the phone call to report back to the base with all our gear. Unfortunately, no one could contact the copilot. He was nowhere to be found. On Sunday, I was able to catch the Steelers' game on TV, and there in the stands was a guy dressed in a gorilla suit wearing a Steelers jersey. Everyone at Barksdale knew who it was, but he never admitted that he was at the game. He had a big smile on his face, however, that didn't disappear for quite some time.

"They Tell You No Lies
When You've Taken To The Skies"
Billy J. Bouquet

Into my life there came a time when I had nothing to do,
So I signed up for four to help them keep score,
Serve my nation with pride
The man said, "Son, we've got a place for you flyin' on a B-52.
We'll send you through school and use you as a tool—
Protecting our American dream."
And they tell you no lies when you've taken to the skies.
Soaring oh, so high.

So you're sittin' in a room, staring at the walls,
Waitin' for the word to come.
'Cause if The Man says, "Hey, you've got to go!"
You know the reason why.
So you jump into your planes, you know you're all insane—
Millions of people will die.
Can you honestly say in the presence of God
You don't have a reason to cry.
It's too late to criticize when you've taken to the skies.
There is no place to hide.

Well, you may come home, you may stay there,
But you'll probably wind up dead,
If you come home, the question I ask, is
"Will you be all alone?"
Get ready my son to see what they've done
With your foolish dreams.
The things you knew no longer exist,
And you know the reason why.
And you can't justify what you've just done
Where are the loved ones you left behind?
Listen to God's cry.

231

"I'm Not Flyin' Tonight"
Billy J. Bouquet

When I went out last night to fly,
The pilot said, "Guns, you sure look high."
I said, "Don't worry, pilot; that ain't all.
Call Fire Control & have the stow pins installed.
Tell the ADO to kiss my butt
The aft compartment door is already shut.

Well, I don't want to fly this mission—
Not with all of that damn transition,
Two A/R's and a pilot pro
To Hell with you and your touch and go's.
When you're flyin' through La Junta on this hot dark night,
I won't be there to clear you left and right.

Talk about nightmares, take the last flight,
When the Radar said, 'There ain't no weather in sight.'
Pardon me, sir, but I think you're wrong.
Climb out of here or you can 'color me gone.'
Tell the EW to call up and close the route;
La Junta Bomb Plot we're climbing out.

Well, here comes the very best part of the flight—
Cel Navigation, I can sleep all night.
Pilot, watch your airspeed—it's startin' to creep;
I guess by now though you're fast asleep.
Center can you give us a little bit more?
Flyin' with the navigator sure is a chore."

Chapter Five

Lest [lest] - *conjunction* - for fear that.

We [wee] - *pronoun* - oneself and another or others.

Forget [for-get] - *verb* - to cease or fail to remember; be unable to recall.

Standing Watch
Doug Cooper

Within humanity are three types of individuals. Sheep are the first and they comprise the largest number. Sheep are nice people, hard working, with families, whose main desires are just to live in peace, raise their families, own homes, and enjoy their friends.

Wolves, on the other hand, are not nice people and only work hard when their own interests are being met. They are often called sociopaths or other scientific and psychological terms. To a wolf, a sheep is only useful as a support system for the wolf's needs. Wolves have self-serving agendas which can be political, religious, or power centered. Wolves have no remorse for their actions. And, for them, no holds are barred.

And then, we have the sheep dogs. Sheep dogs were put on the earth to protect the sheep from the wolves. It's a lonely vocation and one must be in constant motion checking the perimeter and be ready to defend incursions at a moment's notice. Sheep dogs, if given the opportunity, really like to play offense as it usually results in fewer casualties for them and for the sheep they guard. Sheep dogs make great firemen, policemen and, definitely, soldiers, sailors, airmen, and marines.

I was first assigned to SAC at Beale Air Force Base in the spring of 1969. I remember the newcomer's briefing where our commander, Colonel George Burch, stood in front of officers and airmen from all areas of the base and showed a picture of an alert crew responding to a klaxon notification. His concluding statement was that "the alert crews are the reason for this base, for all the activities on the base and for all of your directed duties. If you have a problem with understanding that, come and see me." No one went to see the Colonel.

I was rapidly assigned to a BUFF crew and, after what seemed like a couple of weeks, I assumed alert duties on crew E-30. I had become one of many sheep dogs, standing alert so that others could be secure in their homes and lives. I had joined a fraternity of sailors and airmen who waited in alert shacks, missile silos, or nuclear submarines for the order to go to war to protect the United States.

There were some times when it appeared that we were going to launch and the suspense was as thick as steel. Mostly, however, it involved thousands of hours and days of waiting for the unthinkable to happen. Someone described it as "99 percent boredom and one percent stark terror." But, someone must always "stand the watch."

As most of us know, the alert concept, at least for aircraft, was retired in the last century; however, many men (and women) are still located in forward areas where the comforts we enjoy here at home are seldom available. These folks are the next generation of sheep dogs, standing the watch for us all.

William Whiting wrote the following in 1860:

The Watch
William Whiting

For twenty years,
This sailor has stood the watch

While some of us were in our bunks at night,
This sailor stood the watch

While some of us were in school learning our trade,
This shipmate stood the watch

Yes…even before some of us were born into this world,
This shipmate stood the watch

In those years when the storm clouds of war were seen
Brewing on the horizon of history,
This shipmate stood the watch

Many times he would cast an eye ashore and see his family standing
there,
Needing his guidance and help,
Needing that hand to hold during those hard times,
But he still stood the watch

He stood the watch for twenty years,
He stood the watch so that we, our families,
And our fellow countrymen could sleep soundly in safety,
Each and every night,
Knowing that a sailor stood the watch

Today we are here to say:
"Shipmate…the watch stands relieved.
Relieved by those YOU have trained, guided, and lead
Shipmate you stand relieved…we have the watch!"

"Boatswain…Standby to pipe the side. Shipmate's going Ashore!"

Reflections
Russell Greer

Last night I pulled out *"We Were Crewdogs IV"* and re-read several of the stories. John Cates' story about his last day of alert made me start to reflect on what those years and experiences meant to me. At the end of 1993 I took an early retirement after 15 years to take over the family business when my parents were killed. Now, some 16 years later I have added many more experiences in both business and personal life to my resume. Now I am employed by a major company and travel the country, training associates and giving business advice to many of our franchise owners. I am fairly well compensated and I get to meet and work with many outstanding individuals. In addition, I met and married a beautiful lady in 1998 after my last divorce and she along with the three great children that she already had added to my two wonderful sons meant I had all that anyone could or should need.

With all of that being said I am often transported in my mind through pictures, conversations or phone calls back to my crew days. I realize that those years were the ones that really define my entire life. They represent where I grew up, learned the true meaning of a team, and met individuals that impacted my way of thinking all these years later. In my business we are always trying to get the associates in our stores to work together as a team and in many cases you see them come together and achieve great things. But when I think of real teamwork I think of six men flying together at 500 feet off the ground in the middle of the pitch black night through the mountains of Wyoming, Montana, or Colorado, and knowing that you are in fact part of a team where everyone knows their job and depends upon the others to not just achieve goals but in fact to survive. I think of weeks on alert when we

237

spent more time with each other than most of us ever got to spend with our families at home. We were a family in many ways. I think of deployments to Busy Brewer at RAF Fairford, where we not only flew together but we traveled the country sightseeing in a van. I think of days spent fishing with my aircraft commander, dinners at each other's homes, and watching each other's kids grow up. I think back to Scott Genal, a nav that I flew with at Barksdale that was so full of life that it infected the entire crew. Scott transitioned to B-1s at Dyess and was killed when his crew flew into the side of a mountain in Texas - too young to die.

I think of John Burgess (AC) a pilot that I would fly with anywhere, anytime, and about Eric Single (CP) a great guy but who we always wondered if he was cut out for this life. He's now a Vice-Wing Commander in B-2s so I guess that question was answered. There was Cody Carr (CP), who fell off the roof of his parent's house while on leave in Utah and died after I transferred to Carswell. Mike Ray was a Nav and we would hang out and together burned up a transmission in my truck when we cut and loaded way too much wood into the back. He decided he wanted to go off and fly FB-111s. Bruce Beville was my EW and we made one mean defensive team. Tony Smith, was another EW who was a soft-spoken super-intelligent crewmate that I have lost track of. I remember big Bill Williams, an RN, and a certain little incident on the way to Red Flag that involved a fork and a flying barf bag that went from the black hole to the gunner's seat then on to the pilot's station. There was Mark Turner, an RN, so quite but so good.

These guys are still remembered, along with the many fellow gunners that I have had the honor and privilege of knowing, like Luke "Rooster" Lindsey, the Squadron Gunner in the 62nd BMS when I arrived and just an all around great guy. I can't leave out Marvin Myers, 2nd BMW then CEVG then 8th AF Gunner. I cannot begin to tell you all that I learned from him. Others include Donny Ford, a fellow 62nd gunner and friend, and Jim Howell, later the 62nd BMS Squadron Gunner who suffered a stroke at home and was never to make it back. Tex Ritter was from CEVG and we deployed and roomed together with the 806th BMW (P) at Fairford during Desert Storm. Tom Bettini was also from CEVG, and was very soft-spoken but he knew so much and when I received my S-01 checkride from him the entire flight was a pleasure. CMSGT Lowery, was a SAC Gunner who allowed me to become a part of the 436STS and expand my knowledge and career. There were great guys at the Gunner Training Unit (GTU) at the 436STS at Carswell, like Scott "Skillet" Smith, H-model gunner extraordinaire that I converted into a G-model gunner. Chris Austin passed away in August 2007. Harley Gomes actually became my brother-in-law for seven years, but that is a long story. I also think of John Wing and Pete Gertz, both of whom I have lost track of.

These individuals and so many more are what made those years so special to me. We, along with the many crewmembers who came before us, flew and fought Vietnam and the Cold War, Desert Shield/Storm and the conflicts that followed. We pulled seven-day alert tours so that our families and your's could sleep at night with the knowledge that someone was standing guard. We flew the butt-numbing 10 and 12-hour training sorties - not to mention the 24-hour Chrome Dome missions of the Sixties. These are the ones who made SAC the greatest combat command of all time and gives me a sense of pride that the years have not dimmed. I enjoyed my career after SAC as a Flight Engineer in AMC but there were no integral crews and as such - the feeling was just not the same.

Yes, I enjoy my current career but in the overall scheme of things today we go to work everyday to make money, not to help make the world a safer place. My daily business does not involve life and death business decisions. We will not fly a store into the side of mountain in the dark of night. No one is shooting at us and there are no klaxons to answer, or ORIs, Buy Nones, No-Notice checkrides or EWO study. These days my wife will catch me looking at B-52s in pictures and videos and see a far-away look in my eyes and she does not, cannot, understand. My wife Dianne and I met five years after I retired. She will never know what I am thinking, although she tries. I can only hope that I will have the chance to take her back to Barksdale, back to the 2nd BMW, and show her the BUFFs are still living and still flying. Maybe, just maybe, I can get the Public Affairs Office to give me a flight-line tour and take her inside Boeing Airplane Company's wonderful B-52. What I would give for another eight hours in-flight – just one more sortie.

I hope that one day our kids will understand that what we did was important. We were the Strategic Air Command and "Peace Was Our Profession," and we were very good at what we did.

It was a "Gunner Thing" but it fits all of us…"C'est La Vie."

In the Midst of Ghosts
Tommy Towery

I thought to myself, "Had Charles Dickens written *"A Christmas Carol"* as a war story, then this would have been the setting he would have chosen. I am sure of that. This is the place where Scrooge would meet the three ghosts, had he ever worn a flight suit and ever manned an aircraft."

It was clear to me that among the crowd that was gathered on that day, we had the past, the present, and the future covered. It was a place where many of us were visited by our own ghosts on that warm day. The occasion was not Christmas and the place was not London, but instead it was the occasion of the dedication of the B-52D (Serial Number 0-60683) at Whiteman AFB, Missouri, on July 24, 2009.

The reason for the sun-drenched crowd sitting on chairs and bleachers in front of the static display aircraft on that morning was the dedication of it to the honor and memory of the crew of Ebony 2 - Capt. Robert Morris, Maj. Nutter Wimbrow, 1Lt. Robert Hudson, 1Lt. Duane Paul Vavroch, Capt. Michael LaBeau, and TSgt. James Cook.

The crew being honored did not fly 683 on their Linebacker II mission on the day after Christmas, December 26, 1972. The aircraft they flew, 674, did not return from the hell into which it carried them. It did not survive the three SAM missiles that hit it that night. Neither did the pilot, Capt. Morris, or the EW, Maj. Wimbrow, my former EW school instructor. The other four members of the crew managed to egress the final death spiral of 674 and were captured and held as Prisoners of War until they were released in March of 1973. Their mission that night, along with the other Linebacker II missions, had helped force the North Vietnamese back to the Paris Peace Talks table and ultimately had helped free all the U.S. POWs.

In the row of folding chairs ahead of me sat an obviously proud World War II veteran. He had to be helped to his seat and his arm was in a sling. But when the National Anthem was sung, he did not hesitate to get up and stand as close to attention as his now frail-looking body would allow. I later learned the name and history of this "Ghost of Christmas Past." He was Edward M. Ireland. As a member of the Army Air Corps, SSgt. Ireland served in several positions aboard the B-17 Flying Fortress, including aerial gunner, bombardier, copilot and flight engineer. He was awarded the Distinguished Flying Cross for extraordinary achievement while serving as top turret gunner on a B-17 Flying Fortress on many heavy bombardment missions over enemy occupied continental Europe. In October 1943, he was assigned to the 568th Squadron, Crew 13 in Framingham, England, and flew his first mission on December the 13th of 1943. He had flown two successful bombing missions with his crew, warding off enemy fighter aircraft as an aerial gunner. He was sent to the hospital with influenza and was unable to fly his third scheduled mission. "The last I saw the crew was when the truck left the briefing area," he said. "When I arrived at the British hospital, another flight engineer from another squadron was amazed to see me, because he thought the entire crew was shot down. I spent that Christmas in the isolation ward."

Planes such as the B-17 he flew were still being put on static display at museums and gate guards on bases when the crew of Ebony 2, and many of the rest of us in the audience, were just being trained to fly the mighty B-52., The B-52 Stratofortress had not even flown before his war was over. No one could ever have dreamed of the day that it would be sitting at the entrance of an Air Force base as a museum relic. Many of us present that day embodied the "Ghosts of Christmas Present." We were the ones who now sat in the crowd as the symbol of our youth and our glory days was being dedicated to the next generation. None of us could dare look at SSgt. Ireland and think that one day we too would be his age, if we were lucky. During the dedication and the speeches and presentations, most of my crowd could close our sometimes tear-filled eyes and still see ourselves as young B-52 Crewdogs in Guam and Thailand, in flight suits, and still waiting to be home by Christmas. We were all told it would be over by Christmas - they just didn't say which Christmas. As SSgt. Ireland's aircraft had been dedicated in the past, our aircraft was now being dedicated in the present.

And, behind us on that day, watching the ceremony from afar, stood the "Ghosts of Christmas Future." They were the young ones, the active duty military ones. They were the ones that were now the age we had been when we first earned our wings. They wore flight suits, the same as we had worn, and some had already seen their own combat missions. There was as much difference in us and them as there was in SSgt. Ireland and us. The big difference was that on their suits were similar wings but their patches proclaimed them to be B-2 Crewmembers. Whiteman was an active B-2 base. Yet, in a way, we were all the same. Just as the B-17 crews were retired by the time we started in the B-52, now the B-52 crews that flew during Linebacker II were retired as well. Just as the WWII era B-17s were put out to pasture, our Vietnam era D-model and G-model aircraft had long before stopped flying. Many of them had perished – chopped up in the boneyard in Arizona as victims of the SALT talks and the end of the Cold War. Many of the crewmembers not with us that day were gone as well, existing now only in the memories of those who once knew and flew with them. Of the three generations there, these new young-looking warriors now standing behind us represented the future. Where will they meet their own ghosts of Christmas in their lives?

I found myself wondering, but knew the answer. Does the thought even cross their minds that someday they will be the ones sitting in our place in front of a B-2 that has been retired and was being dedicated as a relic of the past? Do they know that someday they will take their grandchildren to a museum to see the plane they flew in their war? We never thought that way when we were that age, so why should they. Personally, I hope they never even give it a thought. I hope that when they do, I'll have the spirit of SSgt. Ireland and I will be a representative of my generation and sitting on the front row. We need the next generation of warriors to protect us, the way we protected those that came after us, and we were protected by our parents' generation. We don't need them to worry about whether or not they will return from a combat mission.

John Wayne was attributed with saying "Real courage is being scared to death, but saddling up anyway." I don't know if he said that or not, but I like to think he did. That is the way I feel about those that flew during Linebacker II. The crew of Ebony 2 and all the others in their wave knew what awaited them on their mission, but they went, they saddled up anyway. They are the kind of people that I am proud to have associated with during that period of my life. They were the kind of people with whom I now sat at that dedication to other equally brave men.

We all sat there that day with the four survivors and the family and friends of the two that did not survive. We sat in the sun and as the ceremony ended, we looked up, and saw screaming toward us a familiar silhouette. Like the cavalry coming to rescue us at the last moment, the image grew larger and filled our ears with the scream of a ghost of the past, a ghost of the present, and a ghost of the future all rolled into one – a mighty BUFF. The flyover of the B-52 H-model and the audience over which it passed was the inspired reminder of where we came from, where we have arrived, and where we are going.

It served as a reminder to all of us that even thought it is old there is still life left in the B-52 fleet. They still represent the mighty symbol of freedom which served as a common bond to many of us that were assembled on that day. Their memory is a living one and as so serves as a reminder to all of us that we, like it, are not all museum relics yet. I thank God for that every day. I thank God for allowing me to have had a safe landing for each takeoff and to have lived through the days I spent serving my country and fellow Americans as a Crewdog on a B-52. Lest we forget!

Résumé

Résumé [rez-oo-mey] *- noun -* a brief written account of personal, educational, and professional qualifications and experience.

The Contributing Authors

Gerald J. "Jerry" Adler

Captain, USAF (Ret.) earned his commission through the ROTC program at Cornel University in 1953. He began his crew duties on the KC-97 and advanced to become a flight and ground instructor on that aircraft. After attending SOS he transitioned into the B-52 in 1957-1958. He flew them at Castle AFB, California, and was assigned to a combat crew at Loring AFB, Maine, in 1958. He recalls his first copilot was Ken Caldwell. In 1959 he was transferred to Biggs AFB, Texas, where he served three years, ending up on a standboard crew. He moved next to Westover AFB, Massachusetts, and was assigned to a standboard crew there when his accident occurred. Following the accident he spent 14 months in various hospitals and upon his release was offered a choice of a ground job or a medical retirement. Since he had joined the Air Force to fly, he decided he did not want to stay in without that, so he retired in March of 1964.

He attended the University of Houston, earning a law degree in 1966. He earned his Masters in International Law at New York University in 1967, and another Masters from Columbia University in 1969. He was a full time faculty member at University of California Davis for five years and a part-time member for one year before entering into private practice. He worked in local politics for a while and was elected major of Davis. He retired again in 2007 and currently resides in Davis, California. As a result of using his parachute for warmth after his ejection, he became a member of the Caterpillar Club and also earned his membership in the Weber Boosters Club. He has a certificate proclaiming him to be the only person who survived an aircraft ejection without using a parachute.

Scott C. Barbu

Major, USAF Reserves, was in the US Air Force from 1986 to 1992 and then again from 2004 to the present time. He earned his commission from Ohio University AFROTC in 1985, and then earned his navigator wings from Mather AFB in 1986. After graduating from both Electronic Warfare Officer training and Combat Crew Training he was assigned as a B-52G EWO at Andersen AFB, Guam in 1987. The following year he went to B-52H models at K.I. Sawyer AFB, MI, where he upgraded to aircrew instructor and served in Operation Desert Storm in the very last B-52 combat mission of the war, flown from Diego Garcia. He left the Air Force in 1992 and pursued a career in technology sales for 12 years in the St Louis metro area. In 2004 he re-activated his commission into the Missouri Air Guard where he continues to serve today.

Ben Barnard

Colonel USAF, (Ret.) During my Air Force career, my flying assignments took me to Carswell, Blytheville, and Griffiss. Staff assignments included the F-16 System Program Office at Wright-Patterson, the DO's staff at SAC Headquarters, AFROTC duty at Miami University in Ohio, Attaché duty in Mexico City, and Defense Intelligence Agency Headquarters in Washington, D.C. I accumulated over 4,000 flying hours in various models of the B-52 and served as instructor pilot, flight commander, operations officer, and commander of a bomb squadron and commander of an operations support squadron.

After completing 30 years on active duty, my wife of 37 years and I settled in the Hampton Roads area of Virginia. I am more or less still in the military as my post retirement job is senior military analyst with the United States Joint Forces Command.

My wife and I have three adult children and six grandchildren

Billy J. Bouquet

Former Sgt, USAF enlisted in the Air Force in late 1973 at Lackland AFB, TX. After water survival at Homestead AFB, FL and land survival at Fairchild AFB, WA, he completed basic Undergraduate Electronic Training at Castle AFB, CA in December of 1974. Upon completion of that phase of training, he PCS'd to Carswell AFB, TX for CCTS (Combat Crew Training School), which he completed in early 1975. He served as gunner on Crew R/E/S-09 his entire operational time in the Air Force. Following the end of his enlistment in January 1978, he returned to the Houston, TX area to resume his work prior to USAF service. He is now a successful businessman, and he and his wife Jill are happily married and living in Tomball, Texas. They have three children (two daughters and one son).

Doug Cooper

Lt. Colonel, USAF (Ret), got his commission through OTS in 1965 after graduating from the University of U-T ah and facing an imminent first draft pick from the U.S. Army. He completed Undergraduate Navigator Training and Electronic Warfare Training at Mather in 1966 and 1967.

His first assignment was to Beale AFB, and the 744th Bomb Squadron. His crew, E-30, was one of the first selected to attend RTU in D Models at Castle in the summer of 1968 after which he spent six months between the Rock, U-Tapao and Kadena.

After leaving, Beale, he went to Carswell just in time for Bullet Shot and served an additional four TDYs to Guam and Thailand.

After an assignment to the Carswell Command Post (he was the first non-pilot to become a command post controller in SAC), he was reassigned to the SAC Airborne Command Post (Looking Glass). Subsequent assignments at Zaragosa, Spain; Incirlik, Turkey; and Mather AFB were followed by retirement in Sacramento.

Doug and his wife of 39 years, Susan, live in Lincoln, California. His hemorrhoids still ache when he thinks about Giant Lance sorties, of which he had six.

Jack Cotrel

Former SSgt, USAF enlisted in January 1967 as an Aerospace Ground Equipment Repairman. After tech school he was assigned to the 4th Tactical Fighter Wing, 336th Tactical Fighter Squadron, Seymour Johnson AFB, NC. He deployed with 4th Tactical Fighter Wing to Kunson, South Korea, after the Pueblo Incident and served from January 1968 to July 1968. He was assigned to the 8th Tactical Fighter Wing in Jan 1969, Ubon RTAFB, Thailand. He applied for and was accepted for cross training to Defensive Fire Control System Operator/Technician in late summer of 1969. He reported to the 51st Bomb Squadron, Seymour Johnson AFB, NC in January 1970. He completed "Gunnery" School at Castle AFB, CA in August 1970, and returned to Seymour Johnson AFB which was his only permanent stateside duty station. He deployed on Arc Light assignment in January 1972. He served continuous back to back TDY's until April 1973. During his Arc Light tours, he flew approximately 179 missions. His decorations include the Air Medal with 6OLC's, the Air Force Commendation Medal, the Air Force Good Conduct Medal with 1OLC, the Air Force Longevity Medal, the Armed Forces Expeditionary Medal, the Vietnam Service Medal, and the Republic of Vietnam Campaign Medal.

After discharge in June 1973, he attended West Virginia University and East Tennessee State University where he earned an Associate Degree in Law Enforcement and a Bachelor of Science Degree in Criminal Justice. He became as an officer with the ETSU Department of Public Safety in September of 1978. He is presently serving as Associate Vice President for Public Safety at ETSU. He lives in Jonesborough, TN with his wife of 39 years, Jane.

Derek H. "Detch" Detjen

Major, USAF (Ret) grew up in New York City and in the Northern Kentucky/Cincinnati area. As an aeronautical engineering student at the U. of Cincinnati, he flew in one of the first KC-135As during its cold weather testing in Alaska. Entering the Aviation Cadet program at Harlingen, TX in late 1960, he finished as a Distinguished Graduate.. EWO school at Keesler AFB, MS followed where he met his wife Betty. Assigned to Turner AFB, GA, he participated in the first six-month Arc Light tour to Guam in 1966. Returning in both 1967 and 1968 as a crewmember and staff EWO briefer from Columbus AFB, MS, he was part of the first crew to complete 100 Arc Light missions in November of 1967. Crew E-13's 5th Air Medals were presented to them in Guam by the then CINCSAC, General Paul McConnell.

Assigned to the Castle AFB, CA Replacement Training Unit (RTU) in July of 1969, he spent a rewarding four-year tour, training the G and H model crews on Arc Light tactics/ procedures. Later while in 1CEVG, another five-month tour on Guam ensued at Det. 24 "Milky," training the recently deployed B-52 wing. His last five years were spent at Barksdale AFB, LA, in charge of B-52 and KC-135 EWO study, including a trip to RAF Fairford, England during the 1982 Crested Eagle NATO exercise. His military decorations include a Distinguished Flying Cross, eight Air Medals, a Meritorious Service Medal, two Combat Readiness Medals, an Air Force Commendation Medal and several lesser awards.

Major Detjen worked for GD on the Trident submarine at NSB Kings Bay, GA for five years, attained a graduate degree from Valdosta State U. and taught in their on-base education program. A final nine-year stint at Aiken Technical College in SC saw him running their Management and Marketing majors before his retirement in 2000. A lifelong devotee of The Masters, he now lives in Evans, GA, about 10 minutes from the Tournament, which he has managed to attend for over 40 years.

Toki R. Endo

Lt. Colonel, USAF (Ret.) served with the 454th Bomb Wing, Columbus AFB, 320th Bomb Wing, Mather AFB, 93rd Bomb Wing, Castle AFB. Flew both Arc Light and Bullet Shot missions. After retiring from active duty worked on the B-2 as a systems engineer and now works on the B-1 in the same capacity.

Geoff Engels

Captain, USAF (Ret) was a member of the notorious "Red Tag Bastards" - the Class of 1962 from the Air Force Academy. Went to Webb AFB, Big Spring, Texas for pilot training. (Not a "garden spot" but they did have T-38s.) Then got a job as co-pilot on a B-47. When they went to the boneyard, an F-100 assignment appeared, only to be changed to an O-1 upon training completion. After a tour in Nam as a FAC, (513 missions), went to RAF Lakenheath in F-100s. When this tour was up, went to Loring AFB as a B-52G AC, but spent most time in Guam and U-Tapao as a B-52D AC (another 104 missions). Escaped from SAC via the EB-66, only to have the program cancelled as he was handing in his MAC ticket for Korat at Travis. Ended up in Osan, Korea in the O-2 and transitioned to the OV-10. Then went to Hurlburt as an instructor and Chief of the Weapons and Tactics Division. Ended up at Robins AFB as an aerospace engineer (go figure). Retired in 1985 and went to work in Civil Service at Robins. Retired again in 2000 and now works part-time as an aerospace engineer with SSAI. Married for the last 39 years to Sally who, sometimes reluctantly, shared my adventures. Two kids, Nick and Tara who are both engineers. Presently living in Warner Robins, Georgia.

Jim Farmer

Former Captain, USAF was born and raised in Huntington, Long Island, New York. My dad was in the Navy serving in the Pacific during WW II. Graduated form Adelphi University in Garden City NY in 1969, majored in Business. If I didn't do something radical I would have gotten my ass drafted. The only way into the Navy or Air Force as an officer at the time was through their flight/pilot program. I figured in the Navy when you get done flying you're in the middle of the ocean with 5,000 guys, no women and no bar. I joined the Air Force and went into OTS in the fall of '69 followed by pilot training at Reese - Class of 71-05. Wanted to fly big jets, actually liked the idea of a B-52. Assigned to March AFB (So. Cal, here I come).Married my college girlfriend between OTS and UPT. Had four tours to SEA, three as a copilot, one as an A/C and completed 119 ½ missions (got shot down on the 3rd night of Linebacker II. We were the only crew to go down in enemy territory and was rescued - except the RN, Frank Gould. Next was March AFB, California as a BUFF-A/C in the 1974 SAC Bombing and Navigation competition - would have won it too if RN had gone to offset like he was suppose to instead of direct. I served six years, four months, 10 days. I was discharged as a Captain. The airlines were not hiring when I got out so I became a stock broker. I remarried and have two sons. I have been in Seattle the last 33 years, with the last 20 in Bank Brokerage management. The things I miss most about the Air Force are flying and the professionals I got to work with every day. One more thing, I'm still playing competitive hoops.

Russell Greer

MSgt, USAF (Ret.) enlisted in 1977 as an Automated Tracking Radar Systems Technician stationed at Det. 10, 1CEVG, Hastings, Nebraska (actually served with Lt. Col McCrabb another contributor to WWCD). Cross-trained in 1982 to B-52 Gunner. CCTS instructors were Anthony Freeborn and Bruce Helyer; flight line instructor was Mike Riggs. Assigned to the 62nd BMS, 2nd BMW, Barksdale AFB, La., crew R-06. I attended CFIC in 1985 and in 1986 I was assigned to S-03 for the 2nd BMW. In 1987 became S-01 gunner. Participated in 1985 and 1986 SAC Bomb-Nav Competition where our crew finished third in the bombing portion even though we were the only crew still equipped with the ASQ-38. In 1988 I was selected by Headquarters SAC to relocate to the 436STS, Carswell AFB, TX. to become the B-52G Curriculum Development Manager. Completed H Model Qualification and maintained G qualification with monthly TDYs to G units. Deployed with the 806th BMW (Provisional) to RAF Fairford during Desert Storm as a Staff Planner and gunner. Flew numerous combat sorties as lead gunner. B-52G/H, Instructor and Evaluator. When HQ SAC removed the gunners from the B-52 in October 1991 I retrained into Flight Engineer, attended FE training at Altus AFB, Oklahoma and was assigned to C-141s at McGuire AFB, NJ. Took an early retirement in December 1993 after the death of my parents in an aircraft accident to take over the family business in Alabama. Raced the ARCA Superspeedway Series from 1995 to 1997 competing at tracks such as Talladega, Charlotte, Michigan, Daytona and Pocono. Instrument Rated Pilot with over 3000 hours in B-52s, 1000 in C-141s and over 800 as a private pilot, Scuba Diver and a proud father of 3 boys and 2 girls an married to Dianne, the light of my life. Currently reside in Lincolnton, NC. Franchise Field Representative for Aaron's Sales and Lease, traveling coast to coast as a consultant.

Jesman "Jes" A. Hales

Major, USAF (Ret.) I was born at Duke University Hospital and raised in Eastern North Carolina. I received my commission through the AFROTC program and graduated from East Carolina College in 1963. My first assignment was to pilot training in the class of 65-XD at Laughlin AFB, TX where I flew the T-37 and T-33. Upon getting my wings in October 1964, I was assigned to the 325th Bomb Sqdn at Fairchild AFB, WN. While at Fairchild, I had four Arc Light tours. In

1970 I was assigned to the 51st Bomb Sqdn at Seymour Johnson AFB, N.C. While at Seymour, I had one Arc Light augmentee tour and two Bullet Shot tours. I have a total of 300 Arc Light/Bullet Shot missions and 864 days TDY form 1968-1973. From 1973 to 1975, I was an Instructor at Castle AFB, CA. In 1975 I was sent to Air Command and Staff College at Maxwell AFB, AL. I remained on the faculty after my student year. In 1979 I was in charge of the Big Stick War Game. In 1980 I was in charge of the Strategic Forces Curriculum. In 1980 I transferred to the T-39 Detachment at Maxwell and was and instructor in the T-39 until I retired in 1984. Since retirement I have pursued a second career in real estate. My wife and I have three children and two grandchildren. We live on 10 acres in the country near Prattville, Alabama.

Robert O. Harder

Former Captain, USAF An Air Force ROTC Distinguished Military Graduate, Captain Robert O. Harder saw nearly five years of military service, was a Combat Crewman in the Strategic Air Command, and flew 145 combat missions during the Vietnam War (1968-70) as a B-52D navigator-bombardier. His military decorations include the Air Medal with Six Oak Leaf Clusters, Presidential (Air Force) Outstanding Unit Award with Combat "V" Device, National Defense Service Medal, Vietnam Service Medal, Small Arms Marksmanship Ribbon, and the Republic of Vietnam Campaign Medal. He is a commercial pilot and certificated flight instructor.

Mr. Harder received a B.A. degree from the University of Minnesota at Duluth, majoring in Geography and Political Science, with a minor in Air Science. He capped a long retailing career as a vice president of Montgomery Ward & Co., Chicago. Retired from business, he is pursuing a writing career, with an emphasis on military aviation and American frontier history.

Harder and his wife, Dee Dee, live in Chicago and at their summer cabin on Big Sandy Lake, Minnesota. For more information, visit his website at: www.robertoharder.com

Jack C. Hawley

Major, USAF (Ret.) was born and raised in the western suburbs of Chicago. My early teen years surfaced an interest in music and an opportunity to fly in a Cessna 182. That set the aviation "hook". Following H.S. graduation, I was accepted at Southern Illinois University (SIU), to pursue flight through AFROTC and a BS in Marketing.

Freshman year, I "lucked" into a senior ROTC trip to Eglin AFB; setting the hook a bit deeper. The SIU Sport Parachute Club allowed me to jump out of perfectly good airplanes. I applied for USAF pilot training while in Advanced ROTC and with 23 other cadets, completed the Flight Instruction Program, prior to graduation and commissioning. Four of us earned our private pilot's license.

Undergraduate Pilot Training occurred at Reese AFB, where I graduated in absentia. Arlene and I were married that 28 Oct 67, 1,000 miles away. The standard training package for those BUFF-bound, prepped me for initial certification as a copilot at Carswell in Jun of '68.

Four months later our crew was westbound, headed for the Vietnam Arc Light theatre of operation. I participated in six tours (three as copilot, three as aircraft commander) from 1968 through 1973 inclusive. I racked up 380+ combat sorties and 2,100+ combat flight hours. Luckily, I was home, when our first son showed up in Sep '71.

Post-Vietnam assignments included Squadron Officer School, Command Post duty at Carswell and Goose Bay, Labrador (where our second son was born in May of '75). I joined the "rated supplement" as OMS Supervisor at Kincheloe and Seymour Johnson, where I returned to the cockpit as Crewdog/IP. With SAC contingent closure at Seymour Johnson, I returned to Carswell as a Squadron IP, also working out of Training Flight, the Simulator, and sole member of the Instrument School, retiring 31 Oct '87.

Post-military jobs were basically aviation technical writing related; also writing for Nokia, managing a pharmaceutical laboratory, and an 18-wheel semi-, coast-to-coast, driver.

Residing west of Fort Worth for the past 26 years+, married for 41 years+, we're "gramma" and "gramps" to four terrific grandchildren.

Gary Henley

Colonel, USAF (Ret) was commissioned in 1973 through the Texas A&M University ROTC program. His operational flight experience includes the B-52D, RC-135S (Cobra Ball), and RC-135X (Cobra Eye) aircraft, where he served as a flight instructor/evaluator, crew commander, flight commander, operations officer, and squadron commander. He has an extensive background in electronic and information operations (EW & IO) in the areas of training, systems engineering, and defensive systems flight testing. He served in various staff positions at Wing, Center, MAJCOM, and Agency levels in his 30-year military career.

His unique duties included Assistant Deputy Chief of the Central Security Service; Chief, M04 (National Security Agency); and Chairman of the National Emitter Intelligence Subcommittee (under the National Signals Intelligence Committee).

He earned technical specialty badges in both navigation and intelligence career fields, finishing his AF career as the vice wing commander of the 67th Information Operations Wing at Lackland AFB, TX. Currently, he is retired and is a senior technical consultant to the Technical SIGINT Airborne Program Office at the National Security Agency.

Stephen A. Henley

Captain, USAF, was commissioned in 2003 through Texas A&M University's Reserve Officer Training Corps Program. He has active duty flight time in the MH-53, T-1, T-37, T-43, and B-52H. He was initially assigned to Hurlburt Field, Florida as a Communications Officer, where he served two tours in the Middle East at the Combined/Joint Special Operations Air Component Headquarters as the J6/Deputy J6 and the Executive Officer.

After completing an assignment as the 16th Operations Group Executive Officer, he was assigned to the Euro-NATO Joint Jet Pilot

255

Training (ENJJPT) program at Sheppard Air Force Base, Texas. Realizing the unlimited potential of this young aviator, the Air Force reassigned him to Undergraduate Navigator Training (UNT) at Randolph Air Force Base in 2007. After completing Undergraduate Navigator Training with a specialty in Electronic Warfare in 2008, he graduated as a Combat Systems Officer (CSO) and was assigned to FTU (Formal Training Unit) at Barksdale AFB where he graduated in the most recent B-52H class in July 2009. He, his wife Christine, and his two sons (Jonathan and Jackson) currently reside in Minot, North Dakota, where Stephen is assigned to the 23d Bomb Squadron at Minot AFB.

Dave Hofstadter

Colonel, USAF (Ret.) is a Professor at Defense Systems Management College, Defense Acquisition University, Fort Belvoir, Virginia. Prior to this, Dave was a program manager for Battelle Memorial Institute and for Shipley Associates at Tinker Air Force Base, Oklahoma.

On active duty, Dave was Product Group Manager for Information Warfare at Hanscom AFB near Boston, Director of Classified Projects for the B-2 stealth bomber at Wright-Patterson AFB, Ohio, and the B-2 System Support Manager at Tinker. He was also the Program Element Manager for the B-1B bomber at the Pentagon.

Dave accumulated over 2,800 hours as a B-52D electronic warfare officer and flight instructor in the 4018th Combat Crew Training Squadron at Carswell AFB, Texas. He flew 118 combat missions from Guam and Thailand in the B-52D. His decorations include the Legion of Merit, the Distinguished Flying Cross, eight Air Medals and the Republic of Vietnam Gallantry Cross with Palm.

Dave is a native of Macon, Georgia. He has a bachelor of science from the University of Georgia, a master of liberal arts from Texas Christian University, and is a graduate of the Defense Systems Management College and Air War College. Dave and his wife Diane have two children, April and Dan, and live in Centreville, Virginia, where he is a frequent contributor to the We Were Crewdogs series.

Marvin W. Howell

Colonel, USAF (Ret.) was born 1 January 1938 in East Alton, Illinois, and graduated from Southern Illinois University in 1961 as a distinguished ROTC graduate. He entered pilot training at Craig AFB, Alabama in 1961 and subsequently assigned to James Connelly AFB, Texas, where he earned navigator's wings. After completing Electronic Warfare Officer training at Mather AFB, California, in 1963, he went to Glasgow AFB, Montana, where he accumulated over 2,500 hours in the B-52D, including two Arc Light tours and over 100 Arc Light missions. His crew was selected for the Top - 3 award by Third Air Division and was decorated by then General Nuygen Cao Ky during the Guam Summit of Presidents Johnson and Thieu. When Glasgow closed he went to B-52s at Dyess AFB, Texas and a subsequent assignment to EC-47's at Phu Cat AB, Republic of Vietnam in 1969-70, where he flew 118 combat missions in the EC-47.

In 1970, he served on the Squadron Officer's School faculty, and attended Air Command and Staff College before being assigned to the FB-111 wing staff at Pease AFB, New Hampshire. Follow on assignments were to Intelligence Center Pacific, the Joint Strategic Target Planning Staff (JSTPS) at Offutt AFB, Nebraska as Chief of the Data Section in the National Strategic Target List (NSTL) Directorate (1979- 80) and Chief, Combat Targeting Team (1980-81).

Promoted to Colonel in 1981, he worked d as Commander of the Strategic Target Intelligence Center (STIC), Assistant for Air Force Programs, General Defense Intelligence Program (GDIP), Chief of the Weapons Allocation Division in the NSTL Directorate., and Director of Targets, DCS/ In October 1987 he was handpicked to form and direct a new Directorate of Intelligence Systems.

Colonel Howell's awards and decorations include Defense Superior Service Medal, Distinguished Flying Cross, Defense Meritorious Service Medal with One Oak Leaf Cluster, Air Medal with 7 Oak Leaf Clusters, Air Force Commendation Medal with 2 Oak Leaf Clusters, Republic of Vietnam Air Medal, and Republic of Vietnam Air Gallantry Medal.

He retired in September 1990 as the Director of Intelligence Systems at Offutt Air Force Base, Nebraska. He has a master's degree from Troy State University, and a Master's Degree from UNO. Col Howell is married to the former Anita M. Anglin of Roxana, Illinois. They have three children, Bryan and twins Damon and Matthew. He was in private practice as a Mental Health Counselor for five years and is now fully retired.

Thomas D. Jones, PhD

Former Captain, USAF, is a veteran NASA astronaut, scientist, speaker, author, and consultant. He holds a doctorate in planetary sciences, and in more than 11 years with NASA, flew on four space shuttle missions to Earth orbit. On his last flight, Dr. Jones led three spacewalks to install the centerpiece of the International Space Station, the American Destiny laboratory. He has spent fifty-three days working and living in space.

Tom is a Distinguished Graduate of the U.S. Air Force Academy. He piloted B-52D strategic bombers, studied asteroids for NASA, engineered intelligence-gathering systems for the CIA, and helped develop advanced mission concepts to explore the solar system prior to joining NASA's astronaut corps.

Tom's latest title is *Planetology: Unlocking the Secrets of the Solar System* (written with Ellen Stofan, PhD; National Geographic, 2008). *Hell Hawks!* (with Robert F. Dorr; Zenith Press), a true story of an aerial band of brothers in WWII, is the top-selling book at the National Air & Space Museum. The Wall Street Journal named his *Sky Walking: An Astronaut's Memoir* (Smithsonian-Collins, 2006) as one of its "Five Best" books about space. Tom writes frequently for Air & Space Smithsonian, Aerospace America, Popular Mechanics, and American Heritage magazines.

Dr. Jones' awards include the NASA Distinguished Service Medal, four NASA Space Flight Medals, the NASA Exceptional Service Award, the NASA Outstanding Leadership Medal, Phi Beta Kappa, and the Air Force Commendation Medal.

Tom is a member of the NASA Advisory Council, serves on the board of the Association of Space Explorers, and is a regular on-air contributor to television spaceflight coverage. He is currently active in the debate over our nation's space exploration policy.

His website is www.AstronautTomJones.com

Rusty Keller

MSgt, USAF (Ret.) My name is: Harold Keller, but I have always been called Rusty in the military due to having red hair. I was born in 1948 in Torrance, California, but lived most of my growing up years on a cattle ranch east of Bakersfield, California. I entered the Marine Corp after high school, and became an aircraft hydraulics mechanic. I served five years in the Marines and while there flew for almost three years as an F-4 backseater, and during one Viet Nam tour as a CH-46 helicopter gunner. I was discharged from the Marines in late 1970, and less than a year later, I was in the Air Force. I worked as a Crew Chief on KC-135 and B-52 aircraft. I did one Young Tiger TDY, and two Bullet Shot TDY's as an aircraft mechanic. From 1974 to 1982 I was a gunner on B-52 D/F/G/H models, and was an Instructor on the H-model. In SAC, I was stationed at Fairchild AFB, Grand Forks AFB, and Andersen AFB. In 1982 I left SAC for MAC and became a C-130H Flight Engineer at Dyess AFB, until I retired in late 1986 as a Master Sergeant (E-7). While I was stationed at Grand Forks, I took aviation classes at the University of North Dakota, and earned a Commercial Pilot's License, and finished up with two years of college credits (no degree). During 1989/1990 I worked at the B-52G WST simulator in Guam, and during Desert Shield, I helped train the B-52G crewmembers who would eventually fly missions in Iraq during Desert Storm.

"If I had it to do all over again, I'd do it all over again - TWICE!"

Lothar "Nick" Maier

Major, USAF (Ret.) was born in BUFFalo, NY. He entered pilot training with Aviation Cadet Class 55-M, January 1954, and was commissioned at Williams AFB, Arizona, April 1955. Immediately after graduation, he was one of the first Second Lieutenant to enter SAC's Pilot AOB (Aircraft-Observer-Bombardier) course at James Connally AFB, Texas, and received a Navigator rating. Assigned to B-47s at Smoky Hill AFB, Kansas, where in 1956 his crew was the first from the 40th Bomb Wing to be assigned to B-52 upgrade training at

Castle AFB, and subsequently remained there in the 93rd Bomb Wing training cadre.

Nick was a B-52 aircraft commander for twenty years, flying the B through G model aircraft. He received a SAC Crew of the Month Award for an aircraft save in 1967. Served one B-52 Arc Light tour in 1969 with 70 combat missions, and was 8AF Senior Controller at Andersen AFB, Guam, during Linebacker II in 1972. Retiring as a Major in 1977, he worked 16 years in Travel Industry Management. His writings have been published in several military and travel related periodicals. Nick and his wife Mary Beth reside in Texas, and their son Robert is a CFII Rated Commercial Pilot.

Ronald P. Poland

Colonel, USAF (Ret.) was commissioned in the AIR Force in 1973 through Officer Training School (OTS). He has Flight time in B-52G, T-29, and T-43 aircraft. After completing Undergraduate Navigator Training and Electronic Warfare Officer Training in 1974 he was assigned to Barksdale AFB, LA in 1975 where he served as an EW and STAN/EVAL member. In 1979 he was assigned to Air Force EW Intelligence at the Pentagon for an Air Staff Training Assignment. In 1980 he became an EW Instructor at Mather AFB, CA. Following his Mather assignment, he attended Air Command and Staff College at Maxwell AFB, AL. In 1984 he was assigned to Headquarters Air Training Command at Randolph AFB, where he served as a staff officer in charge of all Electronic Warfare training. In 1987 he transferred to Kelly AFB, TX where he served as an Air Force Intelligence Service Detachment commander providing intelligence support to Electronic Warfare. He then attend Air War College in 1992 at Maxwell AFB, AL. Following AWC he was assigned to the Air Force Intelligence Command at Kelly AFB, TX where he provided intelligence support to the Air Force. In 1994 he became director of Operational Support in the Air Force Information Warfare Center (AFIWC). Colonel Poland was then assigned to the National Security Agency, MD in 1995 as Chief of Staff for an operational group. In May of 1997 he retired from the Air Force. Since his retirement, he has been working for CACI as a Director of their San Antonio operations.

Bill Reynolds

(**Major USAF, Ret.**) enlisted in 1964, after graduating from high school in Wilkes-Barre, PA. Following Basic Training, he was assigned to Dyess AFB, TX, where he met and married Yolonda in May of 1966 in the base chapel. Both were 19 years young.

In 1968, Bill was discharged and attended Texas A&M University, graduating in 1971. He entered OTS (SMS, O) in 1972. He completed Navigator Training and NBT at Mather AFB, CA, then CCTS at Carswell AFB, TX, and was assigned to the 20th Bomb Squadron. His first crew assignment in the 20th was as navigator teamed with Jay "Bird" Beasley (Lt Col USAF, Ret). After spending time as a Stan/Eval Nav, crew RN and Instructor, he went to Andersen AFB, Guam. The Reynolds family of five arrived on Guam in August of 1980. There, Bill ended up in the Wing Bomb/Nav shop as Air Weapons Officer, Target Study Officer, and DBNS Project Officer.

In August 1982, after 2,000+ hours in the B-52D, Captain Reynolds jumped the fence to Air Training Command at Mather. There he was a Flight Commander and later, Training Evaluation Branch Chief in the Training Division. While at Mather, Bill completed his Master's Degree in Education.

In 1986, Bill was assigned to HQ ATC at Randolph AFB, TX. He continued his education at Texas A&M accumulating 66 semester hours toward a Doctorate in Adult Education. Unfortunately, "life issues" interfered, and he never completed his degree.

Following retirement on 1/1/1990, Bill worked for various defense contractors until 2003, when he accepted a Civil Service position at Randolph AFB, TX, working with Undergraduate Combat Systems Officer (what USAF now calls Navs and EWOs) training.

Bill and Yolonda make their home in Universal City, TX, and are parents of three grown children and grandparents to five.

Samuel J. Roberts

Lt. Colonel, USAF (Ret.) was commissioned in the Air Force in 1967 through Ball State University Reserve Officer Training Program. He has flight time in the B-52D, B-52H, EB-66C, EB-66C, T-29, and T-43. After completing Undergraduate Navigator Training and Electronic Warfare Officer Training in 1969, he was assigned to K.I. Sawyer AFB, Michigan as a B-52H EWO from 1969 to 1972. He served an ARC Light tour in the B-52D in 1970 (52 Combat Missions) and a Combat tour in the EB-66 (66 combat missions) in 1972 to 73. He was an EW and Navigator Instructor at the Schoolhouse at Mather AFB from 1973 to 78. During that time, he was an EWO Instructor, Nav and EWO Stan/Eval and ITS Instructor. After completing the Armed Forces Staff College in 1979, he was assigned to Air Force EW Intelligence at the Pentagon from 1979 to 83. He also was on the Organization of the Joint Chiefs of Staff from 1983 to 85. Upon completion of Air War College in 1986, he was assigned to Electronic Security Command in San Antonio, Texas and retired in 1988. Since retiring form the military he has worked for various Defense contractors for 7 years and recently retired from Sino Swearingen Aircraft Corp in San Antonio after 12 years. He and his wife, Nancy, live in San Antonio and are enjoying retirement.

He is a Master Navigator with the Organization of the Joint Chiefs of Staff badge. His decorations include the Distinguished Flying Cross, Defense Meritorious Service Medal, Meritorious Service Medal, Air Medal with 5 Oak Leaf Clusters, and the Air Force Commendation Medal.

Kenneth R. "Ken" Schmidt

Major, USAF (Ret.) born in Shattuck, Oklahoma, grew up in Woodward, Oklahoma, and earned his B.S. degree in Business Administration from Southwestern (Oklahoma) State College in 1970. He entered the USAF through Air Force OTS in April, 1974. At Mather AFB, California, he completed Undergraduate Navigator Training (UNT) flying in the T-29 and T-43 aircraft earning his navigator wings and completed Navigator/Bombardier Training (NBT) course in April, 1975. He was assigned to the 9th Bomb Squadron, 7th Bomb Wing, at Carswell AFB, Texas, serving as navigator, instructor navigator, and radar navigator in the B-52D until 1980. His next assignment was the 23rd Bomb Squadron, 5th Bomb Wing at Minot AFB, North Dakota, flying in the B-52H on a combat crew as a radar navigator and instructor, and later as a 5th BW Bomber Scheduler.

In 1983, Ken transferred to Castle AFB, California, as a 4018th CCTS Academic Instructor in the B-52G, and Chief, Air Weapons Branch. In 1987 he moved to Offutt AFB, Nebraska, and the 2nd Airborne Command and Control Squadron (2nd ACCS) as an Operations Plans Officer flying onboard the EC-135C "Looking Glass". Ken was assigned to HQ/SAC DOO and held various staff jobs in the DO community until 1992, then the 55th Strategic Recon Wing, working in the Wing Inspector's Office until retirement in 1993. He accumulated 3,651 flying hours: 155 hours in T-29/T-43 trainers; 2,681 in the B-52D, G, and H; and 811 hours (100 operational missions) in the EC-135C.

Ken completed his Master's Degree from Texas Christian University, and Squadron Officers' School, Air Command and Staff College, and Air War College.

Ken remained in Papillion, Nebraska, after his retirement and is currently an assistant director in the Financial Aid Office at the University of Nebraska at Omaha. Ken was happily married to Susan (Datin) for 34 years and has two daughters, Stephanie and Staci. Unfortunately, Susan passed away on 1 July, 2004, after a five-year battle with breast cancer

George Schryer

SMSgt, USAF (Ret.) grew up in a small town in Ohio and after graduating high school in 1958, joined the US Navy. He spent five years in the Navy, three of which were aboard the destroyer USS Mullinix DD 944 as a Gunners Mate. In 1962 he switched to the Air Force and in 1966 became a Tail Gunner on a B-52 Bomber. He served two tours in South East Asia flying combat missions over South and North Vietnam from Guam, Okinawa and Thailand. In 1977 he was transferred to the US Air Force Survival School in Spokane Washington as a Staff Instructor, and retired in 1980 at the rank of Senior Master Sergeant.

His decorations include the Distinguished Flying Cross, Air Medal with two Oak Leaf Clusters, Purple Heart, Meritorious Service Medal,

Vietnam Campaign Medal with three Stars and the Vietnam Service Ribbon among others.

He is a life member of the Veterans of Foreign Wars, currently holding the position of Commander of the Washington VFW Post 6088 and Commander of VFW District 2 for the State of NC, life member of the Disabled American Veterans, and the Air Force Gunners Assoc. He is also a member of the Vietnam Veterans Assoc. and the Jr. Past Governor of the Washington Moose Lodge and a member of the Beaufort County Committee for Constitutional Studies.

George has been married to Gail E. (Simmons) Schryer for 40 years, and has two daughters, five grandchildren and one great grandchild. He is retired and has lived in "Little" Washington NC since 1994.

Pete Seberger

Major, USAF (Ret.) received an AFROTC commission in June of 1962. He graduated from pilot training at Vance AFB in December of 1963 with class 64-D and was assigned to B-52s at Barksdale AFB. After CCTS, survival, and nuclear weapons schools he went on a crew in June of 1964. His unit was sent TDY to Guam in early 1965 and was transferred to Carswell AFB in May-June of that year. The combined wing then flew most of the early Arc Light missions until December of 1965 In April of 1968 the Carswell unit was reduced to one bomb squadron and Pete went to Grand Forks AFB, upgraded to AC and flew two more RTU tours. In 1972 he was selected as a squadron IPU and in August of 1973 reassigned to Castle AFB as a flight line instructor. In 1977 he went to Clark AB in the 13th AF CAF. He went back to SAC at Ellsworth AFB as a crewdog in 1978, requalified as an AC and IP, and in June of 1979 was selected to command the Physiological Training Unit at Ellsworth, also flying with the 28th Bomb Wing as an attached instructor. In early1983 the Wing was reduced to one bomb squadron so he went on excused status. He retired in December with 6,600 hours in the B-52, some of it every year between 1964 and 1983.

After retiring he worked for a small flight operation in Rapid City and earned his ATP and civilian instructor ratings, then flew a year or so for a regional airline, and finished his flying career as a Flight Safety instructor in the Beechcraft King Air series at the factory school in Wichita. His logbook shows 9,000 total flight hours, over 4,000 as instructor, and 147 combat missions.

David V. "Dave" Thomson

Lt. Colonel, USAF (Ret.) was born in London, Ontario, Canada and grew up in Wisconsin where he completed college and entered the Air Force in 1969. After officer and Navigator flight training, he was assigned to B-52 duties at Carswell AFB at Fort Worth, Texas. His three years in the heavies is recounted in his Crewdog story. In 1975 he left SAC with a new bride and intentions to take a one-year remote tour in AC-130H gunships before returning to SAC. The one year tour ended up being the beginning of 18 years in Special Operations. His duties began with primary flight crew though instructor and evaluator, a tour in Air Force Recruiting, staff duty in operations and plans at all headquarter levels of Special Operations. He concluded his Air Force career in 1993 as Lt Colonel, Chief of Operations Training for all Air Force Special Operations aircrews. Retirement meant more time with his wife, Jana, and two children while teaching elementary school for a couple years. After trying to live on teacher's wages, he got into the oil and gas business as a Petroleum landman. Being a landman has returned him to Fort Worth, currently working for a small independent oil and gas producer.

Tommy Towery

Major, USAF (Ret.) earned a B.S. degree in Journalism in 1968 from Memphis State University where he also earned a commission as a 2nd Lt. through the AFROTC program. Attended Navigator Training and Electronic Warfare Officer Training at Mather AFB, CA. Following B-52 CCTS training in B-52F models at Castle AFB, CA he was assigned to the 20th Bomb Squadron at Carswell AFB, TX flying B-52C and B-52D models. He was deployed for six months to Guam as part of "Operation Bullet Shot" and was assigned to 8th Air Force Bomber Operations as an Arc Light mission planner. On his next deployment he worked as an 43rd Bomb Wing Arc Light planner and flew on B-52 combat missions as a staff officer and helped plan Linebacker II missions. In 1976 his crew received the Mathis Trophy, awarded to the top bomber unit based on combined results in bombing and navigation in the SAC Bomb-Nav Competition.

He logged over 1,600 hours in B-52s and over 5,000 hours total flight time. His decorations include the Meritorious Service Medal with oak leaf cluster, the Air Medal with eight oak leaf clusters, the Armed Forces Expeditionary Service Medal, the Republic of Vietnam Campaign Medal and the South Vietnam Cross of Gallantry with Palm. He retired from the University of Memphis in 2008 and lives in Memphis, Tennessee, with his wife Sue who graciously allows him to devote many hours to his writing hobby. Tommy has written three non-military books, *"A Million Tomorrows –Memories of the Class of '64," "While Our Hearts Were Young," and "Goodbye to Bob."* He was a writer and editor of five other books in the *"We Were Crewdogs"* series and is a member of The American Author's Association and The Military Writers Society of America.

Dave Volker

Former Captain, USAF A North Dakota boy from the start, Dave was born in Grand Forks in 1946. Raised on farms and a few cities in the Midwest, he graduated from the University of North Dakota in 1968. His class standing in UPT, (Class 69-03, Enid AFB) gave him his first choice of aircraft and assignments; a B-52H at Grand Forks, ND. He spent most of his flying career there and did two Arc Light tours accumulating 42 combat missions. While at GFAFB he acquired a second AFSC as the Wing Air Weapons officer.

In late 1974 he was transferred to Blytheville AFB, AR. He assumed the duties of Wing Air Weapons officer there too and moved to the left seat of the B-52G. In early 1975 he was caught in the RIF. He left the USAF in June of that year and moved to the Minneapolis, MN area to pursue a career as a safety professional.

He has worked for several insurance and service related companies over the past 30+ years as an industrial safety consultant. For the past nine years he has focused his work on machine safety and machine controls. He spends his time providing safety advice to a large number of major US corporations including General Electric, Lockheed-Martin, Valero Refining, Rocketdyne, Michelin Tire and Rubber, and 3M. He is particularly proud of his ongoing work at the Johnson Space Center and the shops that support the Astronaut Training Facility there.

Dave is a bachelor and enjoys the occasional company of his sweetie, april z. Together they explore the back roads and byways of Minnesota and Wisconsin on Dave's GoldWing . They are both into RV camping and regularly visit the Experimental Aircraft Association Airventure in Oshkosh, WI. Both are members of the B-52 Stratofortress Association.

<u>PERSONAL NOTES</u>

<u>PERSONAL NOTES</u>